W9-DGE-770

Prentice-Hall Biological Techniques Series
Alexander Hollaender, *Editor*

*Autoradiographic Techniques: Localization of Radioisotopes
in Biological Material*
WILLIAM D. GUDE

Introduction to Research in Ultraviolet Photobiology
JOHN JAGGER

The Laboratory Mouse: Selection and Management
M. L. SIMMONS AND J. O. BRICK

Introduction to Research with Continuous Cultures
H. E. KUBITSCHEK

A Manual of Basic Virological Techniques
GRACE C. ROVOZZO AND CARROLL N. BURKE

Biological Microirradiation: Classical and Laser Sources
MICHAEL W. BERNS

Analytical Ion-Exchange Procedures in Chemistry and Biology
JOSEPH X. KHYM

ANALYTICAL ION-EXCHANGE PROCEDURES IN CHEMISTRY AND BIOLOGY

Theory, Equipment, Techniques

Joseph X. Khym

Biology Division, Oak Ridge National Laboratory
Oak Ridge, Tennessee

209354

Prentice-Hall, Inc., Englewood Cliffs, N. J.

Library of Congress Cataloging in Publication Data

KHYM, JOSEPH X

    Analytical ion-exchange procedures in chemistry and
biology: theory, equipment, techniques.

    Includes bibliographical references.
    1. Ion exchange. 2. Ion exchange chromatography.
I. Title.
QD63.I55K48   541′.3723   74-5343
ISBN 0-13-034942-9

PRENTICE-HALL INTERNATIONAL, INC., London
PRENTICE-HALL OF AUSTRALIA, PTY. LTD., Sydney
PRENTICE-HALL OF CANADA, LTD., Toronto
PRENTICE-HALL OF INDIA PRIVATE LIMITED, New Delhi
PRENTICE-HALL OF JAPAN, INC., TOKYO

10 9 8 7 6 5 4 3 2 1

Printed in the United States of America

CONTENTS

Preface   xi

ONE   Fundamentals of Ion Exchange   1

A. Versatility of Ion-Exchange Materials   1
B. The Chemical Nature of Ion Exchangers   2
C. Chemical Formulas for Ion-Exchange Resins   5
D. Physical Properties of Ion-Exchange Resins   6
   1. Particle Size and Form   6
   2. Swelling and Porosity   8
   3. Crosslinkage   8
E. Chemical Properties of Ion-Exchange Resins   9
   1. Equivalency of Ion-Exchange Reactions   9
   2. Capacity of Exchangers   10
   3. Selectivity of the Resins for the Counter-Ion   10
   4. Stability   12
F. Mechanisms of Ion-Exchange Reactions   12
   1. Ion-Exchange Equilibria   12
   2. The Donnan Membrane Theory   13
   3. Application of the Law of Chemical Equilibrium
      to Ion-Exchange Reactions   17
   4. The Rate of Ion-Exchange Reactions   22
References   23

v

TWO   Ion-Exchange Processes: Batch and Column Techniques;
       Plate Theory of Chromatography   25

       A. Definition and Symbols Used in This Chapter   25
       B. Basic Operations   26
       C. Details of the Batch Method   27
       D. The Column Method   29
          1. General Considerations   29
          2. Frontal Development   30
          3. Displacement Development   31
          4. The Breakthrough Technique   35
          5. Elution Analysis   37
       E. The Plate Theory   42
          1. General Comments   42
          2. The Discontinuous Approach   43
          3. Representation of the Elution Curve as a Gaussian
             Error Function   46
          4. Methods of Determining the Number of Plates in a
             Column   47
          5. The Concept of the Plate Theory as a Continuous
             Process   50
          6. Parameters that Affect the Performance of a Column   53
       References   54

THREE Selecting the Proper Ion-Exchange Material   56

       A. Preliminary Considerations   56
          1. Limitation of Ion-Exchange Materials to Four Classes   56
          2. Selecting the Exchanger by a Process of Elimination   57
             a. Selection based on the net charge of a solute   57
             b. Selection based on the size and net charge of a solute 58
             c. Selection based on the chemical and physical
                environment of a solute   60
       B. The Ion-Exchange Resins   60
          1. Trade Names, Manufacturers, and Suppliers   60
          2. Terminology   61
             a. Particle size   63
             b. Per cent crosslinkage and capacity   64
             c. The ionic form of an exchanger   64
             d. Formulary   65
          3. Chemical and Physical Stability   65
       C. Ion-Exchange Celluloses   67
          1. Fibrous Cellulosic Exchangers   67
          2. Microgranular Cellulosic Exchangers   68
          3. Choice of the Exchanger Type   69
       D. Ion-Exchange Dextrans and Polyacrylamide Gels   71

    1. The Gel Matrix   71
    2. Preparation   72
    3. Some General Characteristics of the Gel-Exchangers   74
    4. Selecting the Type of Gel Exchanger   75
  E. Inorganic Ion Exchangers   76
    1. The Rebirth of Inorganic Ion-Exchange Materials   76
    2. Hydrous Oxide Exchangers   77
    3. Acid Salt Exchangers   78
    4. Heteropoly Acid Salts   79
    5. Metal Sulfides   79
    6. Inorganic Phosphate Gel   80
  F. Miscellaneous Ion-Exchange Materials   80
    1. Some Lesser Used Resinous Exchangers   80
    2. Pellicular Ion Exchangers   81
  G. Ion-Exchange Literature   81
    1. Books   81
    2. Handbook   82
    3. Reviews   83
    4. Journals   83
  References   84

FOUR Laboratory Columns and Accessories:
      Operational Techniques   86

  A. The Chromatographic Assembly   86
  B. Chromatographic Columns   87
    1. Column Shape; Length to Diameter Ratio   87
    2. Column Types   88
    3. Packing the Column   89
      a. Multistage batch-packing   93
      b. Single-stage batch-packing   93
      c. Single-stage pump-packing   93
  C. Line Connections from One Accessory to Another   95
  D. Apparatus for Delivering the Eluent to the Column   96
  E. Detection Devices and Techniques   102
  F. Control of Liquid Flow Through the Column Assembly   104
  G. Fraction Collectors   105
  H. Operation of the Chromatographic Assembly   108
  References   109

FIVE Quantitation of Elution Curves   111

  A. Analysis of Individual Fractions   111
  B. Analysis of Automatically Plotted Elution Curves   112
    1. Methods of Obtaining the Area Under Elution Peaks   112

a. Geometrical integration   113
b. Other lesser used methods   114
2. Conversion of Peak Area to Amount of Solute   115
    a. Direct relation of peak area to sample
        composition   115
    b. Conversion of peak area into weight or moles of
        substance through response factors   115
    c. Conversion of peak area into quantity of substance
        through extinction coefficients   118
C. Constancy of Operating Conditions   119
    1. Preliminary Considerations   119
    2. Internal Standards   119
D. Precision and Accuracy of the Integration Methods   120
E. Resolution   121
References   124

SIX   Simplification of Some Common Analytical
       Chemical Operations   126

A. Nonchromatographic Operations   126
B. Conversion of One Compound to Another   127
C. The Standardization of a Salt Solution by Ion Exchange   128
D. Removal of Interfering Ions   128
E. Preparation of Deionized Water   129
F. Purification of Organic Compounds   130
G. Determination of Iodate Following Periodate Oxidation
    of α-Glycol Groups   131
H. Catalysis   132
 I. Recovery of Trace Constituents   134
    1. Concentration of Trace Substances from
        Water Supplies   135
    2. Recovery of Trace Constituents in General
        Environmental Work   138
    3. The Isolation of Trace Elements in Food Products   138
    4. Separation of Trace Constituents in the Ore and
        Metal Industries   139
    5. Recovery of Solutes from Chromatographic Peaks   140
References   142

SEVEN   Ion-Exchange Chromatography   144

Part I   Ion-Exchange Chromatography of Organic Substances   145

A. Amino Acids, Peptides, and Proteins   145

1. Amino Acids   145
2. Peptides and Proteins   148
   a. Separations on resinous exchangers   148
   b. Separation of peptides and proteins on cellulosic
      ion exchangers   151
B. Separation of Carbohydrate Substances   154
   1. Neutral Carbohydrates   154
   2. Carbohydrate Derivatives   158
C. Organic Acids   159
   1. Introduction   159
   2. Detection Methods   159
   3. Uronic and Aldonic Acids   160
   4. Other Aliphatic Carboxylic Acids   161
   5. Separation of Aromatic Acids   162
D. Amino Sugars   164
E. Amines   165
F. Separation of Cabonyl Compounds in the Presence of
   Bisulfite   168
G. Nucleic Acid Components   168
   1. Brief Historical Account   168
   2. Purine and Pyrimidine Bases and Nucleosides   171
   3. Nucleoside Phosphates   175
   4. Polynucleotides   179
   5. Nucleic Acids   181
H. Simultaneous Analysis of Different Classes of
   Compounds   181

Part II Ion-Exchange Chromatography of Inorganic Substances 182

A. Emergence of Ion-Exchange Elution Chromatography   182
B. Ion Exchange–Complex Ion Interactions   183
C. Classification and Comparison of Chromatographic
   Methods   186
D. Separation of Inorganic Cations   186
   1. Alkali Metals   186
   2. Alkaline Earth Ions   187
   3. Transition Elements and Related Metals   193
   4. The Rare Earth Elements   195
   5. Miscellaneous Metals   197
E. Separation of Inorganic Anions   197
   1. Separation of Chloride, Bromide, and Iodide Ions   197
   2. Chromatography of Polyphosphate Mixtures   199
   3. Separation of Hypophosphite, Phosphite,
      and Phosphate   200
   4. Miscellaneous Other Anions   201
References   201

EIGHT   Utilization of Ion-Exchange Resins for Partition,
        Salting-Out, and Ion-Exclusion Chromatography   208

        A. Preliminary Considerations   208
        B. Partition Chromatography of Carbohydrates   211
        C. Salting-out Chromatography   216
        D. Separation of Ionic Compounds by Ion Exclusion
           and Related Methods   218
        References   223

NINE   Ligand-Exchange Chromatography   224

        A. Description of Ligand Exchange   224
        B. Separation by Ligand-Exchange Chromatography   225
           1. Amphetamine Drugs   225
           2. Purine and Pyrimidine Derivatives   225
           3. Peptides and Amino Acids   228
        References   230

        Appendix A. Extension of the Donnan Principle   235

        Appendix B. Determination of the Total Ion-Exchange
                     Capacity   237

        Appendix C. Determination of the Liquid Volume Held
                     in an Ion-Exchange Column   239

        Index   243

PREFACE

Ion-exchange processes can be used in place of or to simplify many of the chemical operations that routinely confront the experimenter. Employment of these techniques has become so commonplace in the laboratory that the advanced student or practicing investigator not well acquainted with them is at a distinct disadvantage. The separation of like elements into distinct groups or the separation of closely related species within a group, concentration of a trace material, synthesis or degradation of compounds, are only a few of the tasks easily performed with ion-exchange materials. One does not have to be a specialist to practice with assurance the techniques of ion exchange. Ordinary laboratory experience is all the background training that is necessary for one to acquire the necessary skills for experimentations with ion-exchange materials.

One purpose of this book is to encourage those investigators inexperienced in the practice of ion-exchange methodology to adapt these time-saving procedures to many of the chemical manipulations that are carried out in the laboratory. It is intended also to serve as a "refresher" source for those who intermittently utilize ion-exchange methods.

Emphasis is placed on practical experimentation with ion-exchange materials. However, to use these materials effectively, some knowledge of theory is necessary. The amount of theory presented here is kept at a minimum, and what is presented is considered as applying to the ideal case (just as the gas laws apply to an ideal gas). It is not the purpose of this book to render complete historical accounts or give complete bibliographies.

I wish to express my gratitude to Dr. Alexander Hollaender, editor of the Biological Techniques Series, for his encouragement during the preparation of this book. I would like to express appreciation also to Drs. Waldo E. Cohn and James W. Holleman, who painstakingly reviewed the manuscript, and to Dr. J. L. Epler for his valuable comments as to the contents of this text.

I wish to thank the library staff of the Biology Division at Oak Ridge National Laboratory, particularly Azolene Vest, for promptness in acquiring desired literature sources, some very difficult to obtain. Also I am grateful to Nancy Trent who patiently typed and retyped the later chapters of this book.

Finally, my deepest appreciation is expressed to Marylou Khym who typed the rough drafts and ensuing manuscript for the earlier chapters and who in addition handled the vast amount of correspondence that is associated with the preparation of a book.

JOSEPH X. KHYM
Oak Ridge, Tennessee

ANALYTICAL
ION-EXCHANGE PROCEDURES
IN CHEMISTRY AND BIOLOGY

Fundamentals of Ion Exchange

ONE

A. VERSATILITY OF ION-EXCHANGE MATERIALS

Many of the chemical manipulations routinely encountered in the laboratory can be quickly and efficiently carried out by employing ion-exchange materials. Some of these common operations are listed below.

1. Conversion of one salt to another
2. Desalting
3. Concentration
4. Removal of interfering ions prior to analytical determinations
5. Removal of ionic impurities from organic reagents
6. Rapid quantitative determination of ionic solute concentrations (e.g., standardization of analytical reagents)
7. Catalysis
8. Fractionation or separation of both inorganic and organic ions by chromatographic procedures

As implied by their name and as inferred from the list of reactions given, ion-exchange materials react ionically with other ions. Ion exchangers are termed anionic or cationic depending upon whether they take up the negative or the positive ions of a surrounding electrolyte, respectively.

1

B. THE CHEMICAL NATURE OF ION EXCHANGERS

A variety of inorganic and organic substances have been used as ion exchangers. For example, such natural products as proteins, celluloses, carbon, common clays, and various minerals contain mobile ions that will exchange with other ions in a surrounding solution. However, these natural substances have low exchange capacities and other unfavorable chemical and physical properties that limit their practical utilization as ion-exchange substances. Consequently, before 1935, the technique of ion exchange was not widely used as a unit process either in the laboratory or on an industrial scale.

Modern ion-exchange technology began in 1935 with the now classical investigations of Adams and Holmes (1) who discovered that synthetic organic polymers, more commonly referred to as resins, are capable of exchanging ions. These synthetic resins are solids that may be pictured structurally as being composed of two parts. The fundamental framework of these ion-exchange substances is an elastic, three-dimensional hydrocarbon network or matrix; the second part of their structure is hydrophilic in nature and consists of ionizable groups (either acidic or basic) chemically bonded to the hydrocarbon framework. The organic network is fixed, is insoluble in almost all the common solvents used in the laboratory, and is, for all practical purposes, chemically inert. However, the ionizable or functional groups attached to the matrix have active (mobile) ions that can react with or be replaced by other ions. Therefore, if an exchanger particle is brought into contact with an aqueous solution containing ions, the latter may be easily exchanged for those ions initially bound to the resin.

The chemical behavior of an ion-exchange resin is determined by the nature of the functional groups that are attached to the hydrocarbon skeleton. There are two major classes of ion-exchange polymers: *cation exchangers*, whose functional groups can undergo reaction with the cations of a surrounding solution; and *anion exchangers*, whose functional groups can undergo reaction with the anions of a surrounding solution.

A typical cation-exchange resin is prepared by the copolymerization of styrene (I) and divinylbenzene (II). During the polymerization reaction, first linear chains of polystyrene are formed, and these in turn then become covalently bonded to each other, at intermittent points, by divinylbenzene

$$CH{=}CH_2$$

I

$$CH{=}CH_2$$
$$CH{=}CH_2$$

II

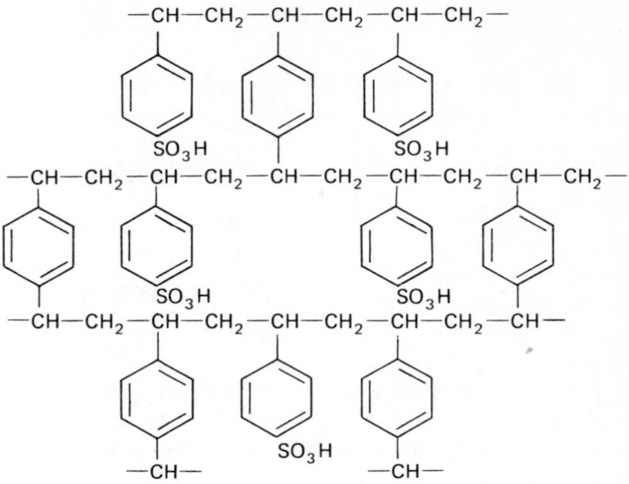

Fig. 1-1. Strong-acid polystyrene type cation-exchange resin.

crosslinks; the result is a three-dimensional insoluble hydrocarbon network. If sulfuric acid is then allowed to react with this copolymer, sulfonic acid groups ($-SO_3^-H^+$) are introduced into most of the benzene rings of the styrene-divinylbenzene polymer, and the final substance formed is a cation-exchange resin (2, 3) whose structure is given in Fig. 1-1.

A typical anion-exchange resin is prepared by first chloromethylating the benzene rings of the three-dimensional styrene-divinylbenzene copolymer to attach $-CH_2Cl$ groups and then causing these to react with a tertiary amine, such as trimethylamine. This gives the chloride salt of a strong-base exchanger (2, 3), which has the structure given in Fig. 1-2.

These crosslinked vinylbenzene resins have remarkable chemical and physical properties. For instance, they are insoluble in concentrated acids, bases, and salts and are resistant to oxidation, reduction, and radiation. The resins have excellent thermal stability and have a high "exchange capacity," which means that a high percentage of the benzene rings of the styrene-divinylbenzene matrix must contain the added ionic functional groups (3). These ionic groups, covalently bonded to the resin matrix, maintain the same chemical properties that they display in aqueous solution; they behave as if they were in the free monomeric form. Consequently, the ionic group fixed to the polymer determines the nature of the ion-exchange material (4). There-fore, just as there are strongly and weakly ionized acids and bases, so can there be these classes of ion-exchange resins. Since there obviously can be many types of ion exchangers, this brings up the question of how best to classify ion-exchange materials.

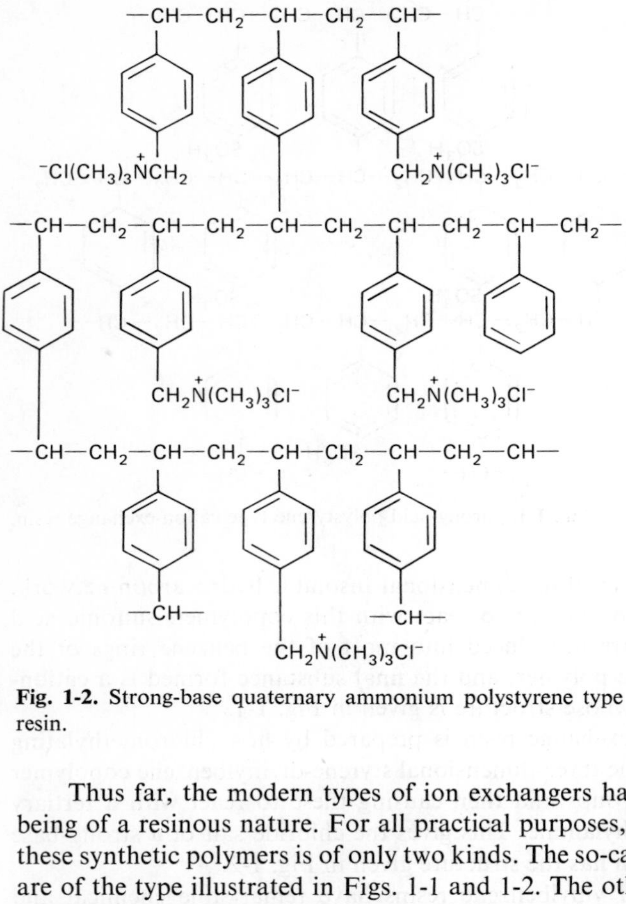

Fig. 1-2. Strong-base quaternary ammonium polystyrene type anion-exchange resin.

Thus far, the modern types of ion exchangers have been considered as being of a resinous nature. For all practical purposes, the lattice material of these synthetic polymers is of only two kinds. The so-called polystyrene resins are of the type illustrated in Figs. 1-1 and 1-2. The other variety of resinous exchanger is prepared by the copolymerization of methacrylic acid (III) and divinylbenzene (II). The result of this reaction gives the weak-acid, acrylic type of ion exchanger (2, 3) that has the structure shown in Fig. 1-3.

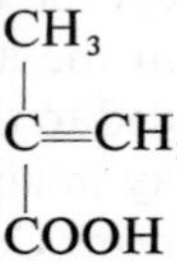

III

In addition to the two kinds of crosslinked vinylbenzene polymers, there are other types of exchange substances of high capacity, such as inorganic ion-exchange crystals and exchange materials made by introducing functional groups into polyacrylamide gels, celluloses, or dextrans. These latter ion-

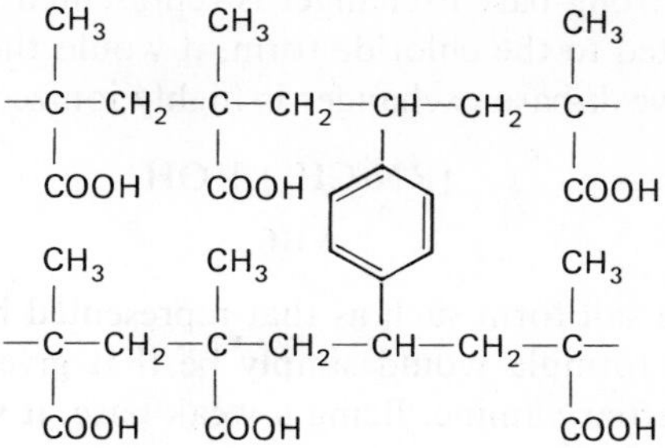

$$
\begin{array}{cccc}
& CH_3 & CH_3 & CH_3 \\
& | & | & | \\
-C-CH_2- & C-CH_2-CH-CH_2- & C- \\
& | & | & | \\
& COOH & COOH & COOH
\end{array}
$$

Fig. 1-3. Weak-acid acrylic type cation-exchange resin.

$$
\begin{array}{cccc}
CH_3 & CH_3 & CH_3 \\
| & | & | \\
-C-CH_2- & C-CH_2-CH-CH_2- & C- \\
| & | & | \\
COOH & COOH & COOH
\end{array}
$$

exchange materials have specialized uses, such as high selectivities for certain ions, or are useful in fractionating macromolecules, such as serum proteins, nucleic acids, and enzymes; these types of exchange materials are considered in a later section (Chapter 3) on "Selecting the Proper Ion-Exchange Material." Many of their properties are given in tabular form in that chapter.

The crosslinked vinylbenzene resins are more versatile and are utilized more than any other exchange material available. For this reason, they will be considered, for the remainder of this book, in more detail than other types of exchangers. Nevertheless, the principles and concepts associated with the synthetic crosslinked resins can readily be applied to other ion-exchange substances.

C. CHEMICAL FORMULAS FOR ION-EXCHANGE RESINS

Ion-exchange resins can be considered as insoluble acids, bases, or salts and, as such, their roles in chemical reactions are easier to visualize if they are assigned chemical formulas. Once the matrix material is specified and is then given a symbol, various kinds of ion exchangers can be classified by formula. Symbols such as ϕ or R are commonly used to represent the lattice material of an ion-exchange substance. By attaching to such symbols the known chemical structure of functional groups, ion-exchange materials are thus conveniently characterized according to their formulas. For example, formulas IV and V represent a weak-acid and a strong-acid cation-exchange resin,

$$RCOOH$$
IV

$$RSO_3{}^-H^+$$
V

respectively. The sodium salts of IV and V, in turn, have the formulas given in VI and VII

$$RCOO^-Na^+$$
VI

$$RSO_3{}^-Na^+$$
VII

A strong-base exchanger is represented by formula VIII, and if it were converted to the chloride form, it would then have the formula denoted by IX.
A weak-base exchanger is highly ionized only when

$$[RN(CH_3)_3]^+OH^- \qquad\qquad [RN(CH_3)_3]^+Cl^-$$

strong base
VIII   IX

in a salt form such as that represented by formula X. In the free base form,
the formula would simply be that given for XI, which is the formula for
a tertiary amine. Being a weak base, it would tend to

$$[RNH(CH_3)_2]^+Cl^- \qquad\qquad RN(CH_3)_2$$

weak base
X   XI

lose its ion-exchange properties at pH values much above 7. It follows then
that ion exchangers have chemical properties similar to those of other ionic
substances. Exchangers with highly ionized functional groups are similar to
strong acids and bases, while those with weakly ionized functional groups
behave like weak acids or bases.

Chemical equations can be written in the usual manner for the reactions
between ion exchangers and other ionic substances. This is illustrated in the
following set of equations.

$$RSO_3^-H^+ + K^+ + Cl^- \rightleftharpoons RSO_3^-K^+ + H^+ + Cl^- \tag{1-1}$$

$$2\,RSO_3^-Na^+ + Ca^{++} + 2\,Cl^- \rightleftharpoons (RSO_3^-)_2Ca^{++} + 2\,Na^+ + 2\,Cl^- \tag{1-2}$$

$$2\,[RN(CH_3)_3]^+Cl^- + 2\,Na^+ + SO_4^=$$
$$\rightleftharpoons [RN(CH_3)_3]_2^+SO_4^- + 2\,Na^+ + 2\,Cl^- \tag{1-3}$$

$$RCOOH + Na^+ + OH^- \rightleftharpoons RCOO^-Na^+ + H_2O \tag{1-4}$$

$$RN(CH_3)_2 + H^+ + Cl^- \rightleftharpoons [RNH(CH_3)_2]^+Cl^- \tag{1-5}$$

D. PHYSICAL PROPERTIES OF ION-EXCHANGE RESINS

1. Particle Size and Form

Most ion-exchange resins are sold in the form of spherical beads. In
a typical preparation, the particles may range from 1 mm to less than 0.04
mm. The resin beads are graded in a range of sizes by manufacturers and
suppliers. The coarser particles (50–100 mesh)* are usually used in batch
operations in which slurry contact of the exchanger and solution is made.
Following the reaction, the resin beads may be separated from the solution
phase by filtration, settling, or centrifugation. The finer resin particles (200–
400 mesh or smaller)* are utilized in chromatographic procedures wherein

*See Chapter 3.

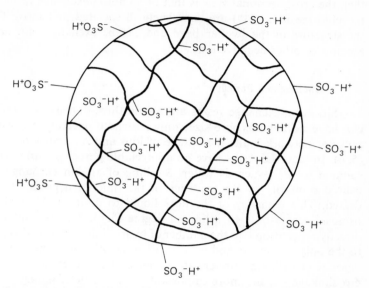

Fig. 1-4. Cation-exchange resin schematic. (Reproduced from R. M. Wheaton and A. H. Seamster in *Kirk-Othmer Encyclopedia of Chemical Technology*, Interscience Publishers, Inc., New York, 1966, Vol. 11, p. 873; by permission of the authors and Interscience Publishers, Inc.)

the resin beads are contained in a tube or column, and solutions to be processed are allowed to flow through the vertical resin bed.

When the beads are immersed in water, they imbibe a limited amount of the liquid to form a homogeneous gel-like structure (5, 6). A cross section of a single cation-exchange bead is schematically illustrated in Fig. 1-4. The wavy lines represent the hydrocarbon network—the solid, insoluble, organic part of the exchanger—to which the ionizable sulfonic acid groups ($-SO_3^-H^+$) are attached. The hydrogen ions (H^+) of this group are completely dissociated in the imbibed water and are free to diffuse throughout the entire resin bead and hence can be exchanged for an equivalent amount of ions of like charge upon contact of the resin bead with a given solution. If the $-SO_3^-H^+$ groups of Fig. 1-4 were replaced by a quaternary amine structure such as XII,

$$CH_3-\overset{\displaystyle \overset{|}{CH_2}}{\underset{\displaystyle \underset{|}{CH_3}}{N^+}}-CH_3Cl^-$$

XII

then the cross-sectional view is that of an anion-exchange resin. In this case the chloride ions (Cl^-) are completely dissociated and move freely within the structure of the swollen bead and can be exchanged for an equivalent amount of other negative ions.

2. Swelling and Porosity

Since ion-exchange resins are elastic three-dimensional polymers, they can have no definite pore size; only a steadily increasing resistance of the polymer network limits the uptake of ions and molecules of increasing size (5, 6). In fact, the resins have no appreciable porosity until swollen in some suitable solvent, such as water. Swelling of an ion exchanger in water is caused primarily by the tendency of the functional groups to become hydrated. This tendency of the pore liquid to dilute itself by the process of osmosis occurs only to a limited degree (5, 6). The amount of swelling is directly proportional to the number of hydrophilic functional groups attached to the polymer matrix and is inversely proportional to the degree of divinylbenzene crosslinking present in the resin. This latter factor—the degree of crosslinkage—is the more important factor in determining to what extent a resin is free to swell or to shrink (7, 8). With a low degree of crosslinkage, the hydrocarbon network is more easily stretched, the swelling is large, and the resin exchanges small ions rapidly and even permits relatively large ions to undergo reaction. Conversely, as the crosslinkage is increased, the hydrocarbon matrix is less resilient, the pores of the resin network are narrowed, the exchange process is slower, and the exchanger increases its tendency to prohibit large ions from entering its structure (3, 7, 9).

This characteristic swelling is a reversible process and at equilibrium there is a balance between two forces. The tendency of the resin particle to undergo further hydration, and hence to increase the osmotic pressure within the bead, is offset by the elastic forces of the hydrocarbon matrix. This osmotic pressure is attributed almost entirely to the hydration of the functional groups, and, since different ions have different degrees of hydration, the particular counter-ion (mobile ion) in an ion exchanger can be expected to have a considerable influence upon the degree of swelling (5, 6, 7).

3. Crosslinkage

The degree of crosslinking in a polystyrene exchanger is expressed as the fraction of divinylbenzene that is contained in the styrene-divinylbenzene resin beads. The content of divinylbenzene varies from 1% to 16% in commercially available resins, with 8% crosslinking being considered ideal for a general purpose ion-exchange resin.

The percentage of crosslinking affects the purely physical structure of

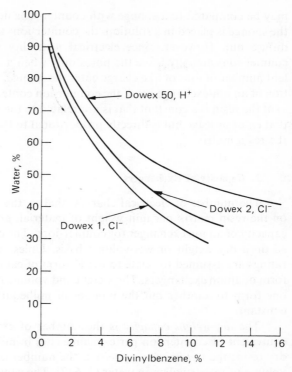

Fig. 1-5. Moisture content vs. crosslinkage. (Reproduced from *Dowex::Ion Exchange*, © 1958, 1959, by permission of the Dow Chemical Company.)

the resin particles. Resins with a low degree of crosslinking can take up a considerable amount of water and swell into a structure that is soft and gelatinous. However, resins with a high divinylbenzene content swell very little; these particles take up only a small amount of water and consequently are somewhat hard and brittle (8). The relationship between moisture content and crosslinkage for some commonly used ion-exchange resins is shown in Fig. 1-5. Dowex 50 is a cátion exchanger of the type shown in Fig. 1-1, and Dowex 1 and 2 are anion exchangers of the type shown in Fig. 1-2.

E. CHEMICAL PROPERTIES OF ION-EXCHANGE RESINS

1. Equivalency of Ion-Exchange Reactions

Ion-exchange reactions progress stoichiometrically; that is, for every ion equivalent removed from a solution by an exchanger, one ion equivalent of like charge must be released. Helfferich (6) suggests that an ion exchanger may be compared to a sponge with counter-ions floating in the pores. When the sponge is placed in a solution, the counter-ions tend to leave the pores and diffuse out. However, since electrical neutrality must be maintained, the counter-ions actually leave the pores only when a stoichiometrically equivalent number of ions of like charge enter the sponge. According to this description of an ion-exchange bead, the counter-ion content—the exchange capacity —of the resin is a constant that is not related to the nature of the counter-ions that enter or leave but is directly proportional to the fixed electrical charge of the resin matrix.

2. Capacity of Exchangers

The number of electrical charges, that is, the number of exchange sites on the resin matrix per unit weight of material, expresses quantitatively the capacity of an ion exchanger for counter-ions. The capacity is usually expressed on a dry-weight or wet-volume basis. Unless otherwise stated, capacity ratings are assumed to relate to the H-form of cation exchangers and the Cl-form of anion exchangers. The weight and volume may change on going from one form to another but the number of milliequivalents (meq) will remain constant.

The *dry-weight capacity* is the number of exchange sites (there is one equivalent of counter-ion per exchange site) in milliequivalents per gram of dry resin; the *wet-volume capacity* is the number of milliequivalents per unit volume of resin swollen in water (3, 6, 7). The capacity defined in this way is a constant that is characteristic of the material and independent of experimental conditions (6, 7).

Capacity ratings are listed by the suppliers of ion-exchange materials. If necessary, capacities may be determined according to the methods given in Appendix B. The capacity of a strong-base or strong-acid exchanger is independent of pH, but the full capacity of a weak-base resin can be utilized only at acid pH values and that of a weak-acid exchanger only in the basic pH ranges. The reason for this is that the former type of exchanger has a low affinity for ions other than OH^- in solutions of high pH and that the latter type of resin shows little attraction for ions other than H^+ in solutions of low pH.

3. Selectivity of the Resins for the Counter-Ion

When an exchanger is added to an electrolyte solution containing a counter-ion different from that initially bound to the resin, an equilibrium is established as illustrated by the cation-exchange reaction:

$$RSO_3^-Na^+ + K^+ + Cl^- \rightleftharpoons RSO_3^-K^+ + Na^+ + Cl^- \qquad (1\text{-}6)$$

The position of the equilibrium depends upon the relative concentration of the competing counter-ions, Na^+ and K^+, both in solution and on the exchanger, assuming equal affinities of the counter-ions for the exchanger. However, in general, and in dilute concentration (< 0.1 N) of electrolyte (8), an ion-exchange resin will show a preference for one ion over another, i.e., the affinities are not equal. Such preferences can be determined for any given series of ions and if they are arranged in a relative order, *selectivity sequences* can be established for the group of ions. For instance, the affinities of some common monovalent cations for a general purpose cation exchanger are in the increasing order*

$$Li^+ < H^+ < Na^+ < NH_4^+ < K^+ < Rb^+ < Cs^+ < Ag^+$$

and for some divalent cations the order is

$$Cu^{++} < Cd^{++} < Ni^{++} < Ca^{++} < Sr^{++} < Pb^{++} < Ba^{++}$$

Likewise for anions, selectivity series have been established. For a general purpose anion-exchanger resin, the sequence in increasing order is

$$F^- < acetate < formate < Cl^- < Br^- < I^- < NO_3^- < SO_4^= < citrate$$

Selectivity sequences depend upon the nature of the functional group (i.e., the fixed ionic group) of the ion exchanger. The discussion so far concerns only strong-acid and strong-base type of resins. For weak-acid resins, H^+ is preferred to any common cation, and for weak-base resins, OH^- is preferred to any of the common anions.

Each ion of a given selectivity scale can be assigned a numerical value relative to a common ion contained in that series. These values are known as *selectivity coefficients*, and the quantitative aspects of such determinations are discussed later in this chapter in a separate section in which the law of mass action is applied to ion-exchange reactions. However, for the present, it will suffice to consider in a qualitative manner some of the factors that influence the position of an ion in a given selectivity series.

Resin selectivity is attributed to many factors. Since ion exchange involves electrostatic forces, selectivity at first glance should depend mainly on the relative charge and the ionic radius of the (hydrated) ions competing for an exchange site (2, 4, 5, 6, 7). For example, different ions of the same valence should have affinities inversely proportional to their hydrated radii; divalent ions should be attracted to an exchanger more than monovalent ions, and trivalent ions should be bonded to a resin more tightly than divalent ions.

---

*Selectivity scales are given in all textbooks on ion exchange. In a later section of this chapter (see pp. 18–21), more specific references on resin selectivity are given.

Factors other than size and charge also contribute to the selection by an ion-exchange resin of one counter-ion in preference to another. Some of these other factors, as discussed by Helfferich (6), Kitchener (5), Kunin (2), Samuelson (7), and Walton (9) are briefly summarized. The extent of sorption increases with:

1. The counter-ion that, in addition to forming a normal ionic bond with the functional group of an exchanger, also interacts through the influence of van der Waals forces with the resin matrix (e.g., ionic compounds with aromatic rings are also held nonionically by resins made from polystyrene).

2. The counter-ion least affected by complex formation with its co-ion or nonexchanging ion (e.g., if the charge on a cation were lowered by the formation of such a complex, its affinity for a cation exchanger would also be lowered).

3. The counter-ion that induces the greater polarization (e.g., Ag^+ has a high polarizability and is more strongly bonded than would be expected from the size of its hydrated radius).

These factors, together with the effect of the size and charge of an ion in exhibiting a certain selectivity toward a resin, are at best only general rules, and as a consequence there are many exceptions to them.

4. Stability

As briefly mentioned above, the resinous ion exchangers are remarkably inert substances; at ambient temperatures, the vinylbenzene-crosslinked resins are essentially insoluble in those solvents and aqueous solutions usually employed in the laboratory. Likewise, at ordinary temperatures and excluding the more potent oxidizing agents, such resins are resistant to decomposition through chemical attack.

Nevertheless, these materials are not indestructible. Extremely powerful oxidizing solutions, such as boiling nitric acid or chromic-nitric acid mixtures cause decomposition. In addition, sulfonic acid resins in the acid form react with water at about 150°C to yield sulfuric acid, and quaternary amine resins in the hydroxide form undergo decomposition above 50°C. Another limitation of these resins is their degradation and degeneration in the presence of strong gamma-ray sources (3).

F. MECHANISMS OF ION-EXCHANGE REACTIONS

1. Ion-Exchange Equilibria

When an ion exchanger is brought into contact with an aqueous solution containing a counter-ion different from that initially bound to the resin, an exchange of ions occurs until equilibrium is achieved. Quantitative descriptions of this exchange process involve well known physical-chemical formulations that fall essentially into two categories. One considers the exchange process as an adsorption phenomenon, and the other categorizes it as the interaction of coulombic forces in electrolytes, such as described by either the mass-action law or the Donnan equilibrium theory.

Adsorption processes described by the Langmuir or Freundlich isotherms seem foreign to interactions concerning ion exchangers and have limited applicability. They are discussed in some detail by Cassidy (10) and by Morris and Morris (11) but are not considered here. On the other hand, considerations of the Donnan equilibrium theory and the mass-action law appear very relevant to the proper understanding of ion-exchange equilibria. Ideality is a condition lacking in most ion-exchange reactions, but the Donnan theory or the mass-action law can always be applied qualitatively and in some instances even quantitatively to ion-exchange equilibria. Before considering these in some detail, it should be beneficial to heed the advice of Kunin (2) and consider whether ion-exchange phenomena should be classified as adsorption or absorption processes. Some interactions can take place on the membrane surface of ion-exchange granules, but most ion-exchange reactions take place within the interior of the resin beads. To avoid any confusion, from this point on, ion-exchange reactions involving the synthetic ion exchangers are referred to simply as *sorption processes* (e.g., the ion A was sorbed to the exchanger by the process of sorption).

2. The Donnan Membrane Theory*

An unusual property, in addition to ion exchange, is displayed when an ion exchanger is immersed in a solution of electrolyte; at equilibrium the concentration of free electrolyte inside the exchanger will be less than that outside. This imbalance in the concentration of electrolyte across the surface boundary of an ion exchanger may be explained by the Donnan membrane theory. This theory applies to the unequal distribution of a diffusible electrolyte between two aqueous phases separated by a membrane permeable to water and to both ions of the electrolyte. For such an imbalance to exist there must be present, in addition to a diffusible electrolyte, a large nondiffusible ion on only one side of the membrane. Because of its physical properties, such a description may be easily applied to an ion-exchange bead (see Fig. 1-4) immersed in an aqueous solution containing an electrolyte. The resin matrix containing the fixed ionic charges constitutes the restricted nondiffusible ion. The boundary surface between the gel-like structure of an ion-exchange bead and the surrounding solution acts as the membrane

*A discussion of the Donnan membrane theory can be found in almost any introductory text of physiology, biochemistry, or physical chemistry; see also the books on ion exchange that are listed in the references for this text.

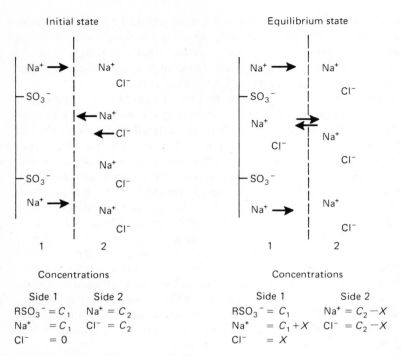

Concentrations   Concentrations

Side 1	Side 2
$RSO_3^- = C_1$	$Na^+ = C_2$
$Na^+ = C_1$	$Cl^- = C_2$
$Cl^- = 0$	

Side 1	Side 2
$RSO_3^- = C_1$	$Na^+ = C_2 - X$
$Na^+ = C_1 + X$	$Cl^- = C_2 - X$
$Cl^- = X$	

Fig. 1-6. Representation of the Donnan equilibrium. The dotted vertical lines denote the surface membrane of an ion-exchange bead that contains fixed negatively charged sulfonate groups attached to the nondiffusible resin matrix R, which is shown as solid lines in the diagram. The negative charge of the sulfonate groups is balanced by the positively charged mobile counter-ions, Na^+ (side 1). Outside the resin particle, Na^+ and Cl^- are always present in equimolar amounts (side 2). The algebraic symbols below each illustrated state represent molar concentrations, so that X moles of NaCl have passed from side 2 to side 1 when equilibrium is established.

(2, 4, 12); water, the counter-ion of the exchanger, and both ions of the electrolyte are permeable to this membrane.

Suppose we place the cation exchanger $RSO_3^-Na^+$ in a dilute solution of NaCl and then allow sufficient time for equilibrium to be reached. The initial and equilibrium states of the system are shown diagrammatically in Fig. 1-6. Since electroneutrality must be maintained, an amount of the Na^+ initially bound to the nondiffusible resin matrix will be retained on side 1. However, on side 2, free, diffusible electrolyte is present and, in response to the concentration gradient, the Na^+ and Cl^- of this side will diffuse, as pairs (again because of neutrality requirements), through the membrane to side 1. After a certain time, an equilibrium will be reached and the rate of diffusion in both directions will be equal. At this point the Donnan equilibrium theory states that

$$[Na^+]_1[Cl^-]_1 = [Na^+]_2[Cl^-]_2 \tag{1-7}$$

where the brackets represent molar concentrations and the subscripts refer to sides 1 and 2 of Fig. 1-6. To satisfy the requirements of electrical neutrality, it is evident that

$$[Na^+]_2 = [Cl^-]_2 \tag{1-8}$$

and that

$$[Na^+]_1 = [Cl^-]_1 + [RSO_3^-]_1 \tag{1-9}$$

From Eqs. (1-7) and (1-8) it can be derived that

$$[Cl^-]_2^2 = [Na^+]_2[Cl^-]_2 = [Na^+]_1[Cl^-]_1 \tag{1-10}$$

and by substituting the value of $[Na^+]_1$ from Eq. (1-9) into the proper bracket of Eq. (1-10), we have

$$[Cl^-]_2^2 = ([Cl^-]_1 + [RSO_3^-]_1)[Cl^-]_1 = [Cl^-]_1^2 + [RSO_3^-][Cl^-]_1 \tag{1-11}$$

which means that

$$[Cl^-]_2 > [Cl^-]_1$$

In other words, the concentration of NaCl is greater on the side of the membrane that is free of the nondiffusible ion.

The inequality in the final concentrations of NaCl may be surprisingly great under certain circumstances. For example, most ion-exchange resins have an exchange capacity or fixed ion concentration that approaches the order of 10 normal. If such an exchanger is immersed in a 0.1 N solution of NaCl, it can be calculated from Eq. (1-7) that the concentration of free NaCl inside the resin bead will be only about 0.001 N. In describing a more unusual case, the Donnan equation also suggests that as the concentration of surrounding electrolyte approaches the fixed ion concentration of the exchanger, the disproportionation of electrolyte concentration across the exchanger-solution interface becomes less and less.

The Donnan membrane theory explains equally well the ability of an anion exchanger to exclude free, diffusible electrolyte. The only difference in this case is that the nondiffusible resin matrix carries a positive charge and the mobile counter-ion is negative.

Although in the foregoing discussion conditions for ion exchange may not be readily apparent, the Donnan theory adequately accounts for the fact that such reactions do take place. That is, even though large amounts of total electrolyte do not penetrate the surface of an exchange resin, the individual ions of the electrolyte may be readily and selectively exchanged for the counter-ions initially sorbed to the resin material; cations selectively penetrate the "membrane" of a cation exchanger and anions migrate through the surface boundary of an anion exchanger. For instance, it can be seen from Fig. 1-6 that if a Na^+ from side 2 enters the interior of a resin bead and at the same time another Na^+ is released from side 1 into the surrounding solution, there is no change in concentrations of the phases at equilibrium. That is, the Na^+ ions are free to wander throughout the entire system with no restraint along the way so long as the condition of electrical neutrality is maintained; thus, if one Na^+ leaves the resin and is released into the solution, there must be another Na^+ that is sorbed to the resin from the solution.

This is also how ion exchange takes place when more than one electrolyte is present in a solution surrounding an ion-exchange bead. Even when several diffusible membrane-penetrating electrolytes are in contact with a resin particle, the Donnan effect still prevails; this is irrespective of the number and nature of other ions that may be present. This aspect of the Donnan equilibrium theory—the interactions of different monovalent ion species and the effect of valency on the interaction of polyvalent ions with ion-exchange polymers—is discussed more fully in Appendix A.

The Donnan principle does not apply to uncharged solutes, hence if small enough and not attracted or repulsed by the matrix material (e.g., through the influence of van der Waals forces or polar interactions), non-electrolytes will diffuse freely in and out of an ion-exchange bead and will tend to have the same concentration both in the resin phase and in the external solution. Thus in these cases, if a sample containing both a nonionic and a dilute ionic solute is percolated through a column containing an exchanger* and then the column is rinsed with a large volume of water, the ionic solute will emerge from the column first, since, due to the Donnan principle, the electrolyte flows around the resin particle but not into it and will emerge when the interstitial volume (the liquid space in between the beads) of the column has been displaced. However, since the nonionic solute diffuses in and out of the resin matrix, it will take a longer time to travel the length of the column and will lag behind the ionic component and emerge from the column last. This phenomenon is known as *ion exclusion* (see 4, 7, 8, 9, 12). Although clearly this technique has nothing directly to do with ion exchange, the phenomenon of ion exclusion enhances the utility of an exchanger in that the total "hold-up" volume (i.e., the liquid between the beads plus the liquid within the beads) of a resin column may be used to effect certain types of separations. Typical examples of the separation of ionic and nonionic substances by ion exclusion are the isolation of salts from ethylene glycol and of hydrochloric acid from acetic acid with water as the rinse solution. When water is replaced by a buffer solution, the separation of either ionic, nonionic,

---

*The counter-ion of the exchanger should be common to one of the ions of the electrolyte, e.g., if NaCl is the ionic solute, the Na-form of a cation exchanger or the Cl-form of an anion exchanger is used. Thus no net exchange of ions occurs.

or mixtures of both can take place within the total liquid volume* of a resin column. Further examples of ion-exclusion chromatography are given in Chapter 8.

The Donnan theory seems to have been "tailor made" to explain how the counter-ions of a particular resin may be exchanged for the ions of like charge in a surrounding electrolyte solution and how, at the same time, co-ions (ions opposite in charge to the counter-ions) are prevented from penetrating the interior of a high capacity ion exchanger. However, excluding the ion-exchange reactions of the low capacity resins in dilute solutions, there are discrepancies between experimental results and calculations made according to a simplified Donnan membrane theory (4, 13). In many respects, the mass-action law more adequately describes ion-exchange equilibria.

3. Application of the Law of Chemical Equilibrium to Ion-Exchange Reactions

It is implicit in Eqs. (1-1) through (1-6) that ion-exchange resins can undergo reversible reactions with other ionic substances. Therefore, the law of mass action is applicable to the state of equilibrium that exists in these reversible reactions. The mathematical form of the law of chemical equilibrium as applied to ion-exchange reactions is expressed in the same manner as it is for any general chemical reaction. But since the activities of ions in the resin phase cannot be evaluated accurately, the equilibrium quotient for an ion-exchange reaction, instead of indicating a true thermodynamic equilibrium constant, is usually expressed as an "apparent exchange constant" or equilibrium quotient that is indicated by the symbol K_Q. Thus for the anion-exchange reaction

$$RN(CH_3)_3{}^+Cl^- + Br^- \rightleftharpoons RN(CH_3)_3{}^+Br^- + Cl^- \tag{1-12}$$

the equilibrium quotient K_Q is calculated from the expression

$$K_Q = \frac{[RN(CH_3)_3{}^+Br^-][Cl^-]}{[RN(CH_3)_3{}^+Cl^-][Br^-]} \tag{1-13}$$

and for the cation-exchange reaction

$$RSO_3{}^-Na^+ + Cs^+ \rightleftharpoons RSO_3{}^-Cs^+ + Na^+ \tag{1-14}$$

the equilibrium quotient is given by

$$K_Q = \frac{[RSO_3{}^-Cs^+][Na^+]}{[RSO_3{}^-Na^+][Cs^+]} \tag{1-15}$$

*The total liquid volume contained in an exchanger bed may be determined by methods given in Appendix C.

In expressing the reaction constant in this manner, the activity coefficients of the ions in both the resin and aqueous phases are usually ignored; the concentrations determined experimentally for a reaction are used uncorrected. In both Eqs. (1-13) and (1-15) the brackets represent the concentration of the reactants and products as expressed in some convenient units. Frequently, normality is the unit chosen for the aqueous phase, while the concentration in the resin phase is usually denoted as equivalents of counter-ion per unit weight of exchanger.

In exchange reactions where the ions are of equal valency, such as shown in Eqs. (1-12) or (1-4) or as in the divalent-divalent reaction

$$(RSO_3^-)_2Ca^{++} + Ba^{++} \rightleftharpoons (RSO_3^-)_2Ba^{++} + Ca^{++} \qquad (1\text{-}16)$$

the value of the equilibrium quotient K_Q is independent of the units chosen to measure the concentrations. However, in exchange reactions between ions of unequal charge, the situation is more complicated due to the differences in the numerical coefficients. These coefficients become the exponents to which the concentration of an ion is raised in the equilibrium quotient. For instance, the equilibrium expression for the reaction

$$3\,RSO_3^-Na^+ + Al^{+++} \rightleftharpoons (RSO_3^-)_3Al^{+++} + 3\,Na^+ \qquad (1\text{-}17)$$

is as follows

$$K_Q = \frac{[(RSO_3^-)_3Al^{+++}][Na^+]^3}{[RSO_3^-Na^+]^3[Al^{+++}]} \qquad (1\text{-}18)$$

The value of K_Q for this latter equation would depend upon the units employed to express the concentrations of Al^{+++} and Na^+ in both the resin phase and in the external solution. In this case the units used will not completely cancel as they do in Eqs. (1-13) or (1-15) or as they do in the equilibrium equations for reactions that involve ions of equal charge.

The equilibrium quotient K_Q of an ion-exchange reaction can undergo a wide variation in value for any given pair of ions; it is constant only at dilute solute concentrations, but otherwise it can vary with the concentration of the involved ions both in the external solution and the amount of each counter-ion bound to the exchanger (4, 6, 12). But even if the K_Q is constant only over a very narrow range of experimental conditions, it is a very useful quantity for expressing ion-exchange data.

As indicated earlier, in the section on selectivity, very seldom is K_Q for an ion-exchange reaction equal to unity. In other words, resins show preference for one ion over another. Therefore, K_Q is a measure of the exchange power of a resin for one particular ion relative to another. In fact, the value of K_Q is often referred to as the *selectivity coefficient* of an ion, and if this index of resin affinity is determined for a series of ions, one then has the data to establish a selectivity scale.

In principle the determination of the order of ions in a selectivity scale is quite simple. For example, the selectivity coefficients shown in Table 1-1 could be obtained in the following manner. Dilute standardized solutions each containing one of the cations listed in Table 1-1 are shaken separately

TABLE 1-1

Relative Selectivity Coefficients of Sulfonic Cation Exchange Resins

Cation	% divinylbenzene (crosslink)		
	4	8	10
Li	1.00	1.00	1.00
H	1.30	1.26	1.45
Na	1.49	1.88	2.23
NH$_4$	1.75	2.22	3.07
K	2.09	2.63	4.15
Rb	2.22	2.89	4.19
Cs	2.37	2.91	4.15
Ag	4.00	7.36	19.4
Tl	5.20	9.66	22.2

Reproduced from Amberlite Ion Exchange Resins—Laboratory Guide, 1964, by permission of the Rohm & Haas Company.

with a weighed amount of resin in a specific cationic state. If the initial state chosen for the sulfonic cation exchanger of Table 1-1 is the H-form, then, in each case, after equilibrium is attained both the resin and solution phases are assayed for hydrogen ion and for the other cation under experiment. These reactions take place according to the equation

$$A^+ + RSO_3^-H^+ \rightleftharpoons RSO_3^-A^+ + H^+ \qquad (1\text{-}19)$$

wherein A^+ equals a cation of Table 1-1. By convention the relative affinity of the resin for A^+ compared to H^+ is given in terms of a selectivity coefficient (or mass-action coefficient) that is defined by the expression

$$K_{H^+}^{A^+} = \frac{[RSO_3^-A^+][H^+]}{[RSO_3^-H^+][A^+]} \qquad (1\text{-}20)$$

The value of each individual $K_{H^+}^{A^+}$ indicates, quantitatively, the preference that the exchanger has for the ion A^+ relative to H^+; if $K_{H^+}^{A^+}$ is greater than unity, the resin prefers A^+ compared to H^+, but if $K_{H^+}^{A^+}$ is less than one, the reverse is true. These selectivity coefficients arranged in increasing numerical order give the selectivity sequences previously presented. However, each time the initial state of the exchanger of Eq. (1-19) is changed (e.g., from $RSO_3^-H^+$ to $RSO_3^-Na^+$), a different set of values is obtained for the selectivity coefficient of the cations; each set gives the same numerical order but has different absolute values. Therefore, for the sake of uniformity, selectivity coefficients for a series of ions are usually given in terms of a scale in which the selectivities of the exchanger for the individual ions of the group are expressed relative to a single chosen reference ion. The selectivity coefficient of the reference ion is arbitrarily assigned a value equal to unity; among cations, Li^+ is often picked as such a standard. Thus by the equation

$$\frac{K_{H^+}^{A^+}}{K_{H^+}^{Li^+}} = K_{Li^+}^{A^+} \qquad (1\text{-}21)$$

Li^+ is the reference ion, and by the previously mentioned convention the value of $K_{Li^+}^{A^+}$ gives the selectivity of the resin for A^+ relative to that of Li^+. The different values for $K_{Li^+}^{A^+}$ give the necessary data to establish the relative selectivity scales for Table 1-1 in which the affinity of the resin for Li^+ is assigned the value of unity. Since Li^+ was chosen as the reference ion in the various cation-exchange reactions involving H^+, the values for H^+ in the table obviously are not calculated by Eq. (1-21). These numbers are obtained by a direct determination of $K_{Li^+}^{H^+}$ for the reverse reaction of Eq. (1-19) or simply by taking the reciprocal of $K_{H^+}^{Li^+}$ for the forward reaction of that equation wherein A^+ equals Li^+.

The selectivity coefficients listed for the anion exchanger of Table 1-2 were determined similarly; here Cl^- was chosen as the common reference ion to compare the relative selectivities of the exchanger for the anions listed in that table.

TABLE 1-2

Relative Selectivity Coefficients of Quaternary Ammonium Anion Exchange Resins

Anion	Coefficient
F	0.09
OH	0.09
Cl	1.0
Br	2.8
NO_3	3.8
I	8.7
ClO_4	10.0

Reproduced from Amberlite Ion Exchange Resins—Laboratory Guide, 1964, by permission of the Rohm & Haas Company.

In actuality the determination of selectivity scales is not as straightforward as it may seem, since the value of a selectivity coefficient can vary with the capacity of a resin, the degree of crosslinking (see Table 1-1), the ionic strength of the equilibration solution, and the percentage of each ion associated with the resin. The effect of these variables in determining the selectivity scale of metal ions toward a cation exchanger is discussed by Bonner and Smith (14); similar data are presented by Gregor *et al.* (15) in their experiments in which the selectivity coefficients of negative ions were determined for anion-exchange resins.

Selectivity scales can be of practical value. For instance, they give the relative displacing power of a series of ions. Knowing the displacing power of one ion relative to another helps one to judge the ease of converting an ion exchanger from one form to another. For example, in converting a known quantity of the exchanger of Table 1-2 from the fluoride form to the chloride form one would estimate that it would require little if any excess of chloride solution that must be put through a column of the fluoride resin, since inspection of Table 1-2 shows that the exchanger has a much greater affinity for chloride than it does for fluoride. On the other hand, the low selectivity of the exchanger for fluoride suggests a large excess of fluoride solution would be needed to convert the resin from the chloride form to the fluoride form. Similarly, selectivity scales give the relative efficiency that ions have as eluting agents (the power to remove sorbed ions from an exchanger). For instance, the data of Table 1-1 indicate that Rb^+ would more easily remove K^+ from a sulfonic acid cation exchanger than would H^+. Selectivity sequences also can be used as a guide to determine how efficient a group of ions will be sorbed to a resin; if the group of ions to be sorbed are positioned below that of the initial counter-ion of the resin in a selectivity scale, then sorption will be more efficient than if the reverse were true.

Since resin preferences give an indication of the separability of a group of ions, it would appear that a very practical application of selectivity coefficients lies in predicting the behavior of ions on an ion-exchange chromatographic column. However, in practice it is much more convenient to predict the column behavior of an ion, for any chosen set of conditions, by employing a much simpler distribution coefficient, D_g, which is defined as the concentration of a solute in the resin phase divided by its concentration in the liquid phase (16, 17), or

$$D_g = \frac{\text{concentration of solute in resin}}{\text{concentration of solute in liquid}}$$

$$= \frac{\%\ \text{M in exchanger}}{\%\ \text{M in solution}} \times \frac{\text{volume of solution}}{\text{mass of exchanger}} \tag{1-22}$$

where M represents the solute under consideration.

As is shown in the next chapter, the value of D_g can be related to the volume of solution (elution volume) required to remove a solute from a chromatographic column that contains a known amount of exchanger. Thus, by determining the D_g for each ion of a group, data is acquired on the practicability of separating these ions on a chromatographic column without the necessity of actually making the column "runs." Such column runs can be very time consuming, since usually a large number of samples are collected and assayed for the presence of a particular solute; also, only one experimental condition can be explored in each column run. However, by batch equilibrations several different samples can be processed simultaneously. Therefore, in a relatively short time, by simply equilibrating small known amounts of resin and solution followed by analysis of the phases, one may easily follow the distribution of several solutes under many different sets of experimental conditions. In this manner one may determine what effect variables such as the capacity and per cent crosslinkage of resin, the type of resin itself, the temperature, and the concentration and pH of electrolyte in the equilibrating solution have on the separability of a group of ions.

By comparing the ratio of the distribution coefficients for a pair of ions, an index as to their separability is obtained. Such a ratio is known as a separation factor* α, which for two ions, A and B, is expressed as (18)

$$\frac{D_g^{A}}{D_g^{B}} = \alpha_{A,B} \tag{1-23}$$

where D_g^{A} and D_g^{B} are the distribution coefficients (see 18), respectively, of ions A and B determined for a specific experimental condition. The more α deviates from unity for a given pair of ions, the easier it will be to separate them on a column of ion-exchange resin.

4. The Rate of Ion-Exchange Reactions

The actual chemical exchange of one ion for another within a resin bead is probably instantaneous (4, 6, 7). However, the rate at which the two ions are brought into contact is controlled by many factors such as the size and charge of the ions involved, the degree of crosslinking of the resin, the temperature of the system, and the size of the resin particles. Much of our present information concerning the rate of migration of ions in and out of a resin bead is due to the work of Boyd and his co-workers (19, 20), who have distinguished between two rate-controlling steps that determine the speed of an ion-exchange reaction: film diffusion and particle diffusion.

Film diffusion pertains to the adherent layer of solution immediately surrounding a resin bead through which ions must penetrate to gain access

*Separation factors are discussed more thoroughly in Chapters 2 and 5.

to the interior of a resin bead. *Particle diffusion* is concerned with the migration of ions within the resin bead itself. In very dilute solutions ($< 0.001\ M$), film diffusion is the rate-determining step controlling ion-exchange reactions, while in more concentrated solutions, as generally prevail in chromatographic columns, particle diffusion controls the rate at which ions exchange (9).

The manner in which the laws of diffusion govern the kinetics of an ion-exchange reaction can be summarized in the following five steps (4, 6):

1. Diffusion of ions through the adherent liquid film surrounding a resin bead.
2. Diffusion of ions within the resin particle to the exchange sites.
3. The actual exchange of one counter-ion for another.
4. Diffusion of the exchanged ions to the surface of the resin bead.
5. Diffusion of the exchanged ions through the adherent surface film into the bulk of the solution.

A rigorous mathematical treatment on the theory of ion-exchange kinetics is given by Helfferich (6, 21). This subject is not further pursued here.

REFERENCES

1. Adams, B. A., and Holmes, E. L., *J. Soc. Chem. Ind.*, **54**, 1 (1935); British Patents 450308-9 (1936).

2. Kunin, R., *Ion Exchange Resins*, John Wiley & Sons, Inc., New York, 1958, 2nd ed., Chaps. 2, 3, 5.

3. Wheaton, R. M., and Seamster, A. H., in *Kirk-Othmer Encyclopedia of Chemical Technology*, Interscience Publishers, Inc., New York, 1966, Vol. 11, pp. 871-97.

4. Berg, E. W., *Physical and Chemical Methods of Separation*, McGraw-Hill Book Company, Inc., New York, 1963, Chaps. 10, 11.

5. Kitchener, J. A., in *Ion Exchangers in Organic and Biochemistry*, Interscience Publishers, Inc., New York, 1957 (ed. by C. Calmon and T. R. E. Kressman), Chap. 2.

6. Helfferich, F., *Ion Exchange*, McGraw-Hill Book Company, Inc., New York, 1962, Chaps. 1, 4, 5, 6.

7. Samuelson, O., *Ion Exchange Separations in Analytical Chemistry*, John Wiley & Sons, Inc., New York, 1963, Chaps. 2, 3, 4.

8. The Dow Chemical Company, *Dowex:: Ion Exchange*, Midland, Mich., 1964 Chap. 1.

9. Walton, H. F., in *Chromatography*, Reinhold Publishing Corporation, New York, 1967, 2nd ed. (ed. by E. Heftmann), Chap. 12.

10. Cassidy, H. G., *Fundamentals of Chromatography*, Interscience Publishers, Inc., New York, 1957, Chap. 9.

11. Morris, C. J. O. R., and Morris, P., *Separation Methods in Biochemistry*, Interscience Publishers, Inc., New York, 1963, Chap. 8.

12. Rieman III, W., and Sargent, R., in *Physical Methods in Chemical Analysis*, Academic Press, Inc., New York, 1961 (ed. by W. G. Berl), Chap. 5.

13. Samuelson, O., *Ion Exchangers in Analytical Chemistry*, John Wiley & Sons, Inc., New York, 1953, Chap. III.

14. Bonner, O. D. and Smith, L. L., *J. Phys. Chem.*, **61**, 326 (1957); **62**, 250 (1958).

15. Gregor, H. P., Belle, J., and Marcus, R. A., *J. Am. Chem. Soc.*, **77**, 2713 (1955).

16. Tompkins, E. R., *Anal. Chem.*, **22**, 1352 (1950).

17. Schubert, J., in *Methods of Biochemical Analysis*, Interscience Publishers, Inc., New York, 1956, Vol. 3 (ed. by D. Glick), p. 252.

18. "Recommendations on Ion-Exchange Nomenclature," *Pure and Applied Chem.*, **29**, 619 (1972).

19. Boyd, G. E., and Soldano, B., *J. Am. Chem. Soc.*, **75**, 6091 (1953).

20. Boyd, G. E., Adamson, A. W., and Myers, L. S., *J. Am. Chem. Soc.*, **69**, 2836 (1947).

21. Helfferich, F., in *Ion Exchange*, Marcel Dekker, Inc., New York, 1966, Vol. 1 (ed. by J. A. Marinsky), Chap. 2.

Ion-Exchange Processes:
Batch and Column Techniques;
Plate Theory of Chromatography

TWO

A. DEFINITIONS AND SYMBOLS USED IN THIS CHAPTER

Most of the terminology used in this chapter is found in the following list. To emphasize or stress a point, other units and symbols are defined as they appear in the body of the text. The first ten definitions are those listed by Kunin (1). With few exceptions, the remainder of the other symbols and notations found in the list and throughout this chapter are those recommended by the International Union of Pure and Applied Chemistry (2). Where recommendations have not been made, the nomenclature in this text closely conforms to that used by Samuelson (3).

Influent = liquid entering ion-exchange column

Effluent = liquid leaving ion-exchange column

Elution = process of removing sorbed ions

Eluent = solution employed for elution = eluant

Eluate = solution resulting from elution

C_0 = concentration of influent

C = concentration of effluent

Concentration history = a plot of C/C_0 versus time or volume of effluent

Saturation column capacity = capacity of column bed when $C = C_0$

25

Breakthrough capacity = capacity utilized up to a predetermined ratio for C/C_0

M = designation for the total amount of a particular solute

M_r = fraction of M in exchanger phase

M_s = fraction of M in external solution

Distribution ratio, D = ratio of the absolute amount of solute in the exchanger phase to the absolute amount in the external solution $(= M \cdot M_r / M \cdot M_s = M_r / M_s)$

Weight distribution coefficient, D_g = the ratio of the total amount of solute per gram of dry ion exchanger to its concentration (total amount per milliliter) in the external solution

Volume distribution coefficient, D_v = the ratio of the total amount of sorbed solute in the ion exchanger calculated per milliliter of column or bed volume to its concentration (total amount per milliliter) in the external solution

P = total number of theoretical plates in a column

$\bar{v}$ = eluate volume at which the solute concentration emerging from the bottom of a column is at a maximum = peak elution volume (ml)

v = eluate volume to any given point (ml)

V_0 = void (interstitial) volume of a column; that volume occupied by liquid between the resin particles

X = a geometric column volume (ml); area of resin bed (cm²) times heights of a bed (cm)

In Fig. 2-1 some of the terms and symbols just defined are represented pictorially.

B. BASIC OPERATIONS

There are two ways of bringing a solution into contact with an ion-exchange resin: the batch method and the column method. In the *batch method*, contact of the exchanger and solution is made by stirring or shaking, i.e., forming a slurry. After equilibrium is attained, the exchanger is separated from the solution phase by filtration, settling, or centrifugation, and both phases are analyzed. The *column method* consists of processing a solution by allowing it or forcing it to flow through a column of resin contained in a tube. Here the solution that emerges from the resin column is collected in consecutive fractions, and these are subsequently analyzed for solute content.

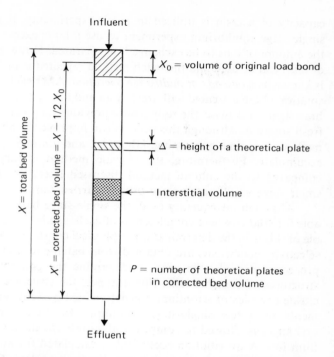

Fig. 2-1. Some terms and symbols associated with the column process. (Reproduced from 0. Samuelson, *Ion Exchange Separations in Analytical Chemistry*, John Wiley & Sons, Inc., New York, 1963, p. 125, slightly modified, by permission of the author and John Wiley & Sons, Inc.)

C. DETAILS OF THE BATCH METHOD

Only in certain instances will it be possible to approach the full capacity of an exchanger in a single-batch procedure. In these favorable cases, the reaction is essentially brought to completion by a driving force such as the formation of weak electrolytes, insoluble products, or stable complexes (4). For example, in the neutralization reaction

$$RSO_3^-H^+ + Na^+ + OH^- \underset{\leftarrow}{\longrightarrow} RSO_3^-Na^+ + H_2O \qquad (2\text{-}1)$$

the formation of water essentially drives the exchange reaction to completion. In the same category, the formation of an insoluble precipitate in the reaction

$$(RSO_3^-)_2Ba^{++} + 2\,Na^+ + SO_4^= \underset{\leftarrow}{\longrightarrow} 2\,RSO_3^-Na^+ + BaSO_4 \qquad (2\text{-}2)$$

forces the exchanger to replace a divalent cation with a monovalent cation. In other types of ion-exchange reactions, such dynamic driving forces are lacking, and in these cases only a small percentage of the total exchange capacity of a resin is utilized in batch experiments. It is impractical in a single-stage equilibrium experiment to use a large excess of resin relative to the amount of ions to be exchanged or separated. However, smaller amounts of resin can be used and the batch method repeated many times; this process is known as a *cascade* or *multistage operation*. After each stage, the separated solution phase is treated with fresh resin and the system is allowed to equilibrate again. Likewise, the resin from a previous stage can be equilibrated with fresh solution. Although this process can improve the efficiency of the batch method, it is time consuming, laborious, and allows experimental errors to accumulate. Furthermore, the cascade method is very inefficient overall compared to the column method discussed later. For these reasons, only single-stage equilibrations are usually carried out.

Although infrequently used, the single-stage batch method is very suitable for studies where completeness of exchange is not necessary. One example of this is in the determination of physical constants. As seen in Chapter 1, selectivity coefficients are obtained from batch-equilibrium data. The batch procedure can also be used to determine the equilibrium constants and structures of complex ions. For instance, the structure and stability of the citrate complex of strontium was determined by Schubert (5) from measurements of a few single-stage equilibria. In these experiments, a cation exchanger is allowed to compete with citrate for the attraction of the strontium ion. A distribution coefficient is calculated for strontium ion both in the presence and absence of citrate. From these data, the dissociation constant and structure of the alkaline earth citrate complex are obtained.

Because of its inefficiency, the batch process *per se* is of little value in separations work. However, it is relatively simple to extrapolate equilibrium data to column operation. It was briefly mentioned in Chapter 1 that the necessary data is acquired by following the partition of solutes between the resin and solution phases by means of a simple weight distribution coefficient D_g (see Eq. 1-22). The equation for calculating D_g is repeated here but is given in the form

$$D_g = \frac{M_r}{M_s} \cdot \frac{v}{m} \tag{2-3}$$

where D_g is the weight distribution coefficient for a solute M equilibrated with m grams of dry resin contained in a solution volume of v milliliters; M_r and M_s are the fractions of solute in the resin and solution phases, respectively. It has been shown experimentally (5, 6) that D_g remains constant over a wide range of resin to liquid ratios (i.e., v/m). This, in turn, is related to an important conclusion that arises from the theory of chromatographic columns (7, 8) and which states that the volume of solution required to elute a solute from a given column of resin can be computed once D_g for that solute is known. Thus, the volume $\bar{v}$ required to elute a solute from a resin column of known dimension, is given by

$$\frac{\bar{v}}{X} = D_g\rho + \epsilon \tag{2-4}$$

where X is a geometric column volume, ρ is the bed density of a column expressed in the units of mass of dry resin per cubic centimeter of column, and ϵ is the void fraction of the column, i.e., the liquid space between the resin particles expressed as a fraction of a geometric column volume, V_0/X. The bed density can be determined in separate experiments by adding a known weight of dry resin to a graduated cylinder containing the eluting solution. After the resin has swelled to its maximum, a direct reading of the settled volume of resin is then recorded (9). It is also a simple matter to determine the void volume of a column (see Appendix C).

Instead of using D_g, some workers prefer to express separation data in terms of a volume distribution coefficient D_v, which is defined as the amount of solute in the exchanger per cubic centimeter of resin bed divided by the amount per cubic centimeter in the liquid phase. The relation between D_g and D_v is given by

$$D_v = D_g\rho \tag{2-5}$$

D_v is most conveniently determined from column experiments; therefore it will be discussed more thoroughly in the section on column methods.

D. THE COLUMN METHOD

1. General Considerations

The column method differs from the batch technique in that, although a single batch of resin is packed into a tube (usually vertical), neither the resin nor the liquid in the column is homogeneously mixed. Let us visualize the column to be divided into a large number of transverse sections of plates (see Fig. 2-1). Then solution flowing into the first section at the top of the column equilibrates with the resin in that section, and the ion-depleted solution that emerges flows into the second plate that contains fresh resin. This process continues all the way down the column and automatically a multistage batch process is carried out (10). In this manner, multiple equilibria are established and, even though the extent of exchange in each plate is limited by the distribution coefficient of the solute, the overall effect is very efficient since a cubic centimeter of resin may contain on the order of 10^{21} exchange sites. Consequently, even solutes that have very low affinities for an exchanger can be quantitatively removed. Accordingly, the column method is used where complete exchange is necessary or when it is required to separate, with maximum efficiency, a group of ions having very similar exchange potentials. There are three fundamental types of column operations: *frontal, displacement*, and *elution analysis*. Of the three techniques mentioned, the latter is the most widely used and is the most effective method for separation work. Consequently, concepts and principles of elution analysis have been more thoroughly explored than those of frontal and displacement analysis.

2. Frontal Development

This is the simplest form of chromatography and consists of continuously passing a sample through the column of resin. The rate at which the various components move down the column depends upon their relative selectivity coefficients. The component of the mixture with the lowest coefficient will move the fastest. A schematic illustration of this technique, taken from Morris and Morris (11), is shown in Fig. 2-2. A, B, and C are three components of a mixture that is fed continuously to a column. As the experiment progresses, the positions of the fronts will be those shown in Fig. 2-2(a). The least-sorbed component, A, of the sample mixture has moved fastest, and it is followed, in order, by the slower moving B, and lastly by C, which has the greatest affinity for the exchanger and therefore has progressed the least. Pure A can be collected until B reaches the bottom of the column, as indicated in Fig. 2-2(b). Subsequently, A and B appear together in the column eluate until the C front emerges. After this point, as seen from Fig. 2-2(c), the composition of the eluate is identical to that of the feed solution. Since only a portion of the fastest moving component can be separated, frontal

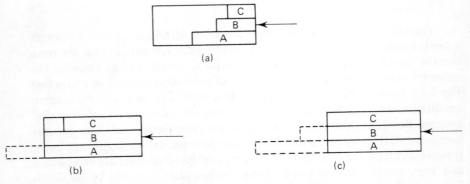

(a)

(b)   (c)

Fig. 2-2. Schematic illustration of frontal development: (a) solute bands develop, those with higher affinities move slower; (b) solute B reaches bottom of column; (c) point at which all three solutes appear in effluent at the same concentration as they have in influent. (Reproduced from C. J. O. R. Morris and P. Morris, *Separation Methods in Biochemistry*, Interscience Publishers, Inc., New York, 1963, p. 12, by permission of the authors and Interscience Publishers, Inc.)

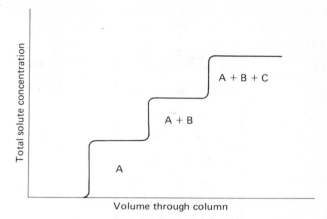

Fig. 2-3. Chromatogram of three solutes (A, B, C) by frontal analysis.

analysis is of limited use for quantitative analysis. It can be used to estimate the minimum number of components in a mixture. Each increase in concentration as components emerge from the bottom of the column represents a solute initially present in the feed solution. Thus, if the concentration of solute in the eluate is plotted against the volume of sample put through the column, one obtains the type of chromatogram shown in Fig. 2-3. Each step in the elution curve represents at least one solute present in the sample.

3. Displacement Development

Reichenberg (10), and Rieman and Sargent (12) give exceptionally clear accounts of this type of chromatography and the examples they used to illustrate this technique are also given here.

The essential requirement for this process is that the ions of the sample have a higher affinity than the ion initially bound to the exchanger. Let us suppose that a solution of sodium chloride is passed continuously through a bed of the cation exchanger $RSO_3^-H^+$, which is initially surrounded only by water. The requirement for displacement is satisfied since the relative selectivity coefficient of the sodium ion compared to hydrogen ion, for this exchanger, is greater than unity (i.e. K_H^{Na} is > 1). A convenient way of demonstrating displacement development by the so-called *breakthrough technique* is schematically illustrated in Fig. 2-4. When the sodium chloride solution makes contact with the top layer of the cation exchanger, sodium ions are taken up and hydrogen ions are released. This results in the appearance of hydrochloric acid in the interstitial volume surrounding the resin beads. As the sodium chloride solution moves down through the column, successive fresh layers of resin are encountered and further exchanges take place in

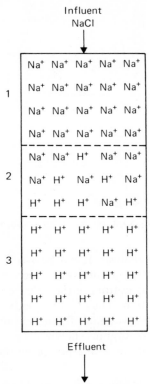

Influent
NaCl

Effluent

Order of appearance: H₂O, HCl, NaCl

Fig. 2-4. Conversion of a cation exchanger, initially surrounded only by water, from the H-form to the Na-form.

which additional sodium ions are sorbed to the exchanger and equivalent amount of hydrogen ions are released into the solution phase. Before all of the initial interstitial volume of water has been flushed from the column, three zones will have developed; these are indicated in Fig. 2-4 as 1, 2, 3. The middle zone, 2, where all the exchanging is occurring, contains both sodium chloride and hydrochloric acid in the liquid phase, and both hydrogen and sodium ions bound to the resin. Exchanger in the top zone, 1, has been completely converted to the Na-form and the sodium chloride solution now passes unchanged through this portion of the exchanger. The bottom zone, 3, still contains the exchanger in its initial H-form, and the hydrochloric acid produced as the column continuously removes sodium ions passes unchanged through this section of the column. Finally, all of the interstitial volume of the column is washed out and then hydrochloric acid emerges from the bottom of the column and appears in the effluent. Its concentration quickly approaches that of the incoming sodium chloride solution and remains at

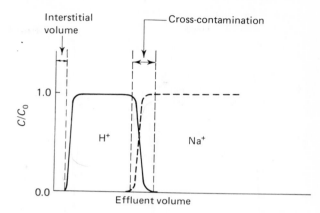

Fig. 2-5. Chromatogram for the displacement of H^+ by Na^+: exchanger initially in H-form surrounded only by water; C = ionic concentration in effluent; C_0 = ionic concentration of Na^+ in influent. (Reproduced from W. Rieman III and R. Sargent, in *Physical Methods in Chemical Analysis*, W. G. Berl (ed.), Academic Press, Inc., New York, 1963, p. 154, slightly modified, by permission of the authors and Academic Press, Inc.)

this concentration until the exchange zone reaches the bottom of the column and sodium ions appear in the column eluate; this is called the *breakthrough point* for the sodium chloride solution. The concentration of the sodium ions rapidly approaches the influent concentration and the hydrogen-ion concentration quickly falls to zero. If this process is followed quantitatively, the data can be represented by a displacement chromatogram such as shown in Fig. 2-5. Displacement can be used to analyze a mixture of ions. Suppose we allow a small volume of solution containing sodium and potassium ions to flow through a column of $RSO_3^-H^+$ and this, in turn, is followed by a large volume of cesium chloride solution. Soon a series of sharp bands will emerge from the column of resin. Hydrogen ion appears in the effluent first, followed by sodium, potassium, and cesium in turn; they all appear in the effluent at the same concentration as that of the incoming, displacing cesium ions. The order of their removal from the column is in the order of their increasing selectivity coefficients. This type of separation is illustrated in Fig. 2-6. Characteristic of displacement analysis is the sharp boundary between the bands as they emerge from the column. Nevertheless, since the bands are contiguous, there is always cross-contamination in the boundary regions so that an entire band does not emerge 100% pure. However one important feature of displacement chromatography is that up to one-third of the bed capacity may be used (12) in analysis or purification work.

An ion of lower affinity can dislodge an ion having a higher selectivity

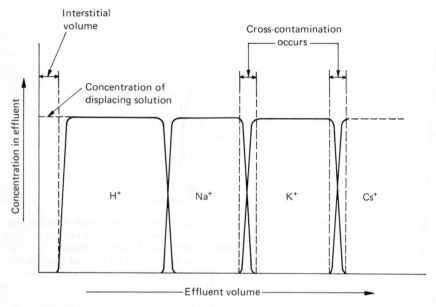

Fig. 2-6. Hypothetical separation of Na$^+$ and K$^+$ by displacement chromatography: cation exchanger initially in H-form; elution zones were developed with CsCl solution. (Reproduced from D. Reichenberg, in *Ion Exchangers in Organic and Biochemistry*, C. Calmon and T.R.E. Kressman (eds.), Interscience Publishers, Inc., New York, 1957, p. 96, slightly modified, by permission of the author and Interscience Publishers, Inc.)

coefficient from a resin column. For instance the sodium-hydrogen exchange of Fig. 2-4 can be reversed, whereupon hydrogen ion replaces sodium ion. However, the exchange process is no longer classified as displacement. One reason is that we have the inversion in the selectivity coefficients (i.e., K_{Na}^H is < 1), which is opposed to the requirement of displacement that states the eluate ion must have the higher selectivity coefficient. Another reason is that the boundary lines of the exchanging zone (section 2 of Fig. 2-4) no longer have the sharpness that is a characteristic of displacement. In fact, in the case of the hydrogen-sodium exchange, the middle zone is now wider and it increases in width as it passes down the column (10). Both the front and rear boundaries are blurred and they become more and more diffuse as the now ill-defined exchanging zone moves down the column. This type of column behavior, where an ion of lower affinity removes one of higher affinity and creates "self-diffusing" boundaries, is known as *elution*, not displacement (10). Before describing elution analysis in some detail, further discussion and uses of the breakthrough technique seem to be in order.

4. The Breakthrough Technique

The replacement of one ion for another by either displacement or elution is conveniently followed by the breakthrough method described for Fig. 2-4. Furthermore, capacity measurements of the resin bed are easily determined by this technique. Breakthrough data is acquired by plotting the volume of solution that is passed through a column versus the concentration of the "breakthrough ion" as it appears in the effluent. Thus, in the case of the H-Na exchange, fractions are collected and titrated with standard alkali to determine the concentration of H^+ present in each. When the concentration of H^+ emerging from the bottom of the column is equal to that entering the top of the resin bed, the experiment is terminated. If C_0 is the known concentration of H^+ entering the exchanger bed and C is the concentration of H^+ that emerges, then a plot of C/C_0 versus the volume of effluent gives a typical breakthrough curve such as is shown in Fig. 2-7. In most cases a symmetrical curve results and the theoretical capacity of the exchanger bed is obtained from volume b, which corresponds to the point $C/C_0 = 0.5$. The breakthrough capacity of the column is represented by volume a. The breakthrough capacity is a quantity that is not well defined and is dependent on several factors. Under ideal conditions the slope of the straight portion of the sigmoid breakthrough curve depends upon the relative affinities of the two involved ions. When the exchange takes place by displacement, the linear

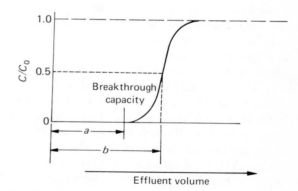

Fig. 2-7. Breakthrough curve for H^+ that results from converting the Na-form of a cation exchange to the H-form: C = concentration of HCl in the effluent; C_0 = concentration of HCl in the influent; milliequivalents represented by volume b is the theoretical capacity of the exchanger; milliequivalents represented by volume a is the breakthrough capacity, or operating capacity, of the exchanger. (Reproduced from E. W. Berg, *Physical and Chemical Methods of Separation*, McGraw-Hill Book Company, Inc., New York, 1963, p. 195, by permission of the author and McGraw-Hill Book Company, Inc.)

portion of the breakthrough curve is more vertical, hence the breakthrough capacity approaches the value for the theoretical capacity. On the other hand, when an ion of lower affinity is used to displace an ion of higher exchange potential (i.e., elution) the reverse is true and a flatter sigmoid breakthrough curve is obtained; this widens the difference between the theoretical and breakthrough capacities. Other factors that affect the breakthrough capacity of an exchanger bed are the particle size of the resin beads, the flow rate at which influent passes through the column, the temperature, the composition of the solution, and the column dimensions. These factors are discussed on both a theoretical and practical basis by Samuelson (3) and are not considered in detail here.

In practice, for analytical work, an amount of exchanger must be chosen so that breakthrough does not occur while a sample is still being admitted to the resin bed; otherwise the uptake of the sample ions will not be quantitative. If some prearranged value of breakthrough is chosen, say $C/C_0 = 0.001$, then this means that 99% of the solute is removed from the last volume fraction collected.

Another practical feature of the breakthrough technique is that it is readily amenable to simple and rapid methods for the determination of the volume distribution coefficient D_v. These methods, as shown by Kraus and Nelson (13), can be used to decide the optimum conditions for chromatographic separation. Let us suppose we want to devise a chromatographic scheme to separate sodium and potassium on a strong-base cation exchanger using hydrochloric acid as the eluent. The cation exchanger in its H-form is slurried into two columns (for such exploratory experiments the column need only be a few centimeters high and a few millimeters in diameter). The columns are then equilibrated with 0.5 M hydrochloric acid solution which we will assume is chosen as the eluting solution. Next a solution of sodium ion, say, at a concentration of 0.001 M in 0.5 M hydrochloric acid is passed through one of the columns; a solution containing potassium ion (also 0.001 M in 0.5 M hydrochloric acid) is percolated through the other column. Flow is continued through both columns until the alkali metal concentration in the effluent equals that of the influent (i.e., until complete breakthrough occurs). The breakthrough curve for these experiments is typical of the one shown in Fig. 2-8. From volume b corresponding to the breakthrough point $C/C_0 = 0.5$, the amount of sodium or potassium ion retained per cubic centimeter of the exchanger bed X can be calculated, it being equal to $b(C_0/X)$; C_0 is the concentration of alkali metal ion in the external solution (the interstitial volume) after complete breakthrough of each ion has occurred. By definition

$$D_v = \frac{\text{amount of sorbed solute per cm}^3 \text{ of bed volume}}{\text{amount of solute per cm}^3 \text{ of external solution}} \qquad (2\text{-}6)$$

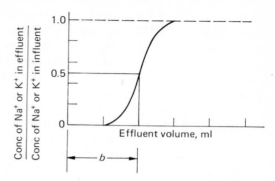

Fig. 2-8. Breakthrough curve that determines the capacity of a cation exchanger for Na⁺ or K⁺ under a hypothetical set of conditions (see text for details).

Thus for the case where 0.5 *M* hydrochloric acid is the eluting reagent, the D_v's for sodium and potassium are easily computed. If the ratio of the volume distribution coefficient of sodium to potassium is very close to unity, indicating incomplete separation, the experiment is repeated using a lower concentration of hydrochloric acid. Once separation* is indicated, the volume required to separate sodium and potassium on a single column of any dimension using hydrochloric acid as the eluent, can be calculated from

$$\bar{v} = (D_v + \epsilon)X \tag{2-7}$$

where the quantities of Eq. 2-7 are those previously defined.

5. Elution Analysis

A distinctive feature of this method is that all the components of the sample can be completely resolved and quantitatively recovered. The chief drawback to the method is that only small quantities of solute can be analyzed. As a general rule, the quantity of sample sorbed to the exchanger in elution chromatography should be less than 5% of the bed capacity. Preferably the sample is sorbed to the resin column from a small volume of solution, of low eluent ion concentration (giving high distribution coefficients) so that initially the solutes occupy only a narrow band at the top of the column. Next, a solution of higher ionic strength is run through the column to move the solute bands down through the exchanger bed. During this movement, the composition of the sorbed sample continuously changes, since ions with

*The degree of separation as indicated by the ratio of D_v's can be given in terms of a resolution factor. This manner of expressing separation data—quantitatively—is explored more thoroughly in Chapter 5.

lower affinities move fastest down the column while those with higher relative selectivity coefficients are retarded. Eventually the components of the original solute band are resolved into single independent zones that contain only one solute. The process of elution for a two-component mixture is illustrated schematically in Fig. 2-9. For the case of very similar ionic species, one band may immediately follow another, and sometimes the two may even overlap. However, most often the bands are separated by a region (*l* of Fig. 2-9) in which the only solute present in the solvent flowing through that section of the column are the eluting ions. Furthermore, since the eluting ions are more weakly sorbed than the ions being separated, they will always be present in excess during the entire course of the elution and will precede the appearance of the solute bands in the column effluent. By way of analogy, Walton (14)

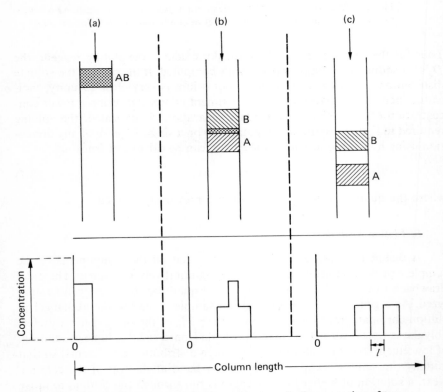

Fig. 2-9. Elution analysis: (a), (b), (c) successive stages in the separation of two solutes A and B. The distance represented by *l* is a region in which the only solute present in solvent flowing through that section of the column are the eluting ions. (Reproduced from C. J. O. R. Morris and P. Morris, *Separation Methods in Biochemistry*, Interscience Publishers, Inc., New York, 1963, p. 14, slightly modified, by permission of the authors and Interscience Publishers, Inc.)

likens elution to rolling a heavy rock down a stream bed by passing many times its weight of water over it.

It was mentioned earlier that during elution, i.e., when an ion is replaced by an ion of lower exchange potential, self-diffusing boundaries arise at both the leading and trailing edges of the exchange zone as it moves down a column. This means that the concentration gradient within each solute band is not sharp. In fact, the concentration of solute increases to a maximum at one point in the exchanging zone and decreases on the other side of the point. The zone widens as it moves down the column and the maximum concentration steadily decreases. Therefore elution graphs do not actually appear as those given in Fig. 2-9, but are bell-shaped and they take the form of the Gaussian curve of error. Such curves for a pair of hypothetical solutes are shown in Fig. 2-10, in which the volume of eluting solution passed through the column is plotted against the concentration of the solutes as they appear in the effluent.

The effluent volume, $\bar{v}_A$ or $\bar{v}_B$ of Fig. 2-10, at which the solute concentration reaches its maximum, is an important constant of elution analysis. The quantity is known as the *peak elution volume* (or *ml to peak*) and it determines the chromatographic properties of a solute under a given set of conditions; once known for one column, the ml to peak is readily calculated for columns of other dimensions. In terms of the number of column volumes (i.e., $\bar{v}/X$) the peak elution volume is essentially equal to the volume distribution coefficient D_v for large values of $\bar{v}$ compared to X; for smaller values of $\bar{v}$ it is readily computed. This can be seen in the following derivation set forth by

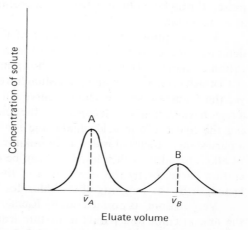

Fig. 2-10. Gaussian-shaped elution curves illustrating the separation of two solutes, A and B, by elution analysis. $\bar{v}_A$ and $\bar{v}_B$ represent the volume of eluate at which the solute concentrations, respectively, reach their maximum.

Kraus and Nelson (13). They computed D_v from the following relationships:

$$D_v = \frac{v}{Ad} - \epsilon = V_d l - \epsilon \qquad (2\text{-}8)$$

$$D_v = V_{max} - \epsilon \qquad (2\text{-}9)$$

Here v is the volume of eluent (cm³) that moves a solute band a distance d (cm) down a column of length l (cm) and of cross-sectional area A (cm²) and which has an interstitial fractional volume of ϵ; V_d is the same volume of eluent expressed in geometric column volumes (i.e., $l \cdot A$, which in our terminology is equal to X). When the solute band moves off the column, d becomes equal to l, V_d becomes equal to V_{max}, the number of column volumes of effluent to the elution maximum, and Eq. 2-8 reduces to Eq. 2-9, which written in our terminology, becomes

$$D_v = \frac{\bar{v}}{X} - \epsilon \qquad (2\text{-}10)$$

This, in turn, is a transposed form of Eq. 2-7, presented previously in connection with the calculation of D_v from breakthrough data. The fractional interstitial volume ϵ (that fraction of the column occupied by liquid between the resin beads) is about 0.4 for most columns. This is a close approximation to the theoretical value for the hexagonal close packing of uniform spheres (i.e., the resin beads). In most cases, the first term of Eq. 2-10 predominates; thus D_v is essentially equal to $\bar{v}/X$, the number of column volumes that have passed through a column when the concentration of a solute in the eluate is at a maximum.

The exceptional resolving power of the elution method is proficiently demonstrated by the separation of sodium and potassium on a strong-acid cation exchanger. This separation, shown in Fig. 2-11 is taken from the work of Cornish (15). Sodium and potassium have very similar chemical properties, yet the Gaussion-shaped elution curves of Fig. 2-11 are well separated by a region where neither ion appears in the column effluent. By simply evaporating the hydrochloric acid eluate, the ions are quantitatively recovered in a pure state as their chlorides. If an unknown sample is analyzed, the amount of alkali metal in each solute band can be obtained from the area under the elution curve, or by the summation of the normality of each fraction collected, or by the weighing of the chlorides after drying.

Very seldom, as pointed out by Rieman and Sargent (12, also see 16) will the first attempt to separate a mixture lead to ideal elution peaks. Usually by trial and error method, a series of chromatographic separations such as those represented by the elution graph of Fig. 2-12 will be encountered. Poor resolution, ill-defined or oddly shaped elution curves, or time wasted in elut-

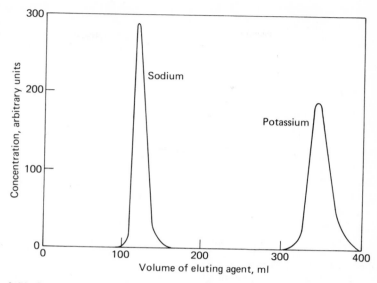

Fig. 2-11. Separation of 9 mg of potassium from 3 mg of sodium on a strong-acid cation exchanger in the H-form: column size 20 cm in height, 0.9 cm in diameter; eluting solution 0.6 *M* HCl at a flow rate of about 1 ml per min per sq cm at room temperature. (Reproduced from F. W. Cornish, *Analyst*, **83**, 634 (1958). by permission of the author and the Society for Analytical Chemistry.)

ing peaks far apart are results of poor operational techniques that can be corrected. The number of variables that affect the resolution efficiency are many. Some of these are: the concentration and nature of the eluent, the flow rate, the column dimensions, and the particle size of the exchanger. If the best set of conditions is to be found by trial and error, a considerable number of time consuming experiments must be carried out. However, theoretical treatments have been developed by which optimum conditions can be predicted for difficult separations. In general, mathematical treatments of column theory are exceedingly complicated. The approach known as the *plate theory* gives a more clear-cut picture of the column process and is simpler mathematically than some of the other theoretical treatments. Furthermore, the complex equations of plate theory can be reduced to simple terms that, on a practical basis, are readily used to predict such quantities as zone shapes, peak position, and the length of column needed for a given separation. The plate theory of elution chromatography is the only theoretical approach considered here.

Two different viewpoints are presented. In the Mayer-Tompkins (7) approach, liquid flow through a column is visualized as a discontinuous process by which resin and solution are equilibrated step by step in successive

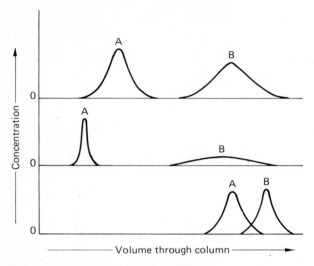

Concentration

Volume through column ⟶

Fig. 2-12. Three types of elution curves that are often encountered by the trial and error method of separating two solutes, A and B. (Reproduced from W. Rieman III, *Record Chem. Progress*, **15**, 85 (1954), slightly modified, by permission of the author and the Kresge-Hooker Scientific Library.)

horizontal sections (called plates) of a column (see Section D). A more precise description of solvent flow through an ion-exchange column is given by Glueckauf (8), who treats equilibrations of resin and solution phases in each segment of the column as a continuous process.

E. THE PLATE THEORY

1. General Comments

The plate theory as applied to chromatography is derived from the mathematical treatment given to the successive equilibrations in a fractional distillation column. The same concept of measuring the efficiency of a distillation column by finding the number of theoretical plates in the column can also be applied to measuring the resolving power of an ion-exchange column. A *plate* is a theoretical transverse layer of cross-sectional area equal to the area of the column. The height equivalent of a theoretical plate (HETP) given by

$$\text{HETP} = \frac{\text{length of column}}{\text{total number of plates in a column}} \tag{2-11}$$

The plate concept was first adapted to chromatography by Martin and

Synge (17) who developed the theory to predict solute behavior on liquid-liquid partition columns. It was later extended to ion-exchange columns by Mayer and Tompkins (7) and given a more thorough treatment by Glueckauf (8).

In deriving the equations of the plate theory, it is assumed (3, 7, 8, 11, 12, 14) that: (1) the solution phase reaches equilibrium with the exchanger of each plate before it flows into the next plate; (2) the distribution of solute between the resin and solution phases is the same in any plate, i.e., as the solute moves down the column it is always distributed in constant proportion; (3) the concentration of the solute being analyzed, both in the resin phase and in the solution phase, is negligible compared to the concentration of the eluting ions present in both phases.

2. The Discontinuous Approach

The following descriptions and derivations of the discontinuous approach of the plate theory are condensed versions of those given by Mayer and Tompkins (7), Morris and Morris (11), and Walton (14). The important conclusions that emerge from this theory will be better understood if the relationships of the following definitions and notations of symbols are kept clearly in mind. The column is considered as consisting of a large number of theoretical plates P, each containing an equivalent mass of resin m and volume of solution, Δv. As flow is started, these Δv's of solution move down the column from plate to plate.

$V_0 = P\,\Delta v =$ volume of solution in the column (a "column volume")

$N\ =$ number of V_0's that have entered a plate up to a given moment

$F\ =$ number of V_0's passed through column

$\quad = \dfrac{\text{total volume through column}}{V_0} = \dfrac{\Delta v N}{P\,\Delta v} = \dfrac{N}{P}$

$D\ =$ distribution ratio $= \dfrac{\text{amount of solute in exchanger (at equilibrium)}}{\text{amount of solute in solution (at equilibrium)}}$

Under the conditions of the plate theory, as the solute M moves down a column, the amount of solute found in the resin and solution phases throughout the entire column is governed by

$$D = \frac{M_r}{M_s} \tag{2-12}$$

Therefore when a unit quantity of solute is introduced into the initial segment or plate (plate 0) of a column and allowed to equilibrate, the fraction of solute in the solution phase is $1/D + 1$ and that in the resin phase is D/D

+ 1. (See Fig. 2-13, where r represents the resin phase and s the solution phase and Σ is the total fraction of material in any plate P.) Now if the solution of plate 0 moves into plate 1 (the next adjacent plate) and fresh solvent now containing only the eluting ions moves into plate 0 and the system is allowed to equilibrate again, the situation is that shown in Fig. 2-13(c). The distribution of solute after three such transfers and equilibrations is that given in Fig. 2-13(f). After the process is repeated many times, the solute, which originally occupied a narrow band at the top of the column, is now spread out in a Gaussian distribution with a maximum concentration at one point in the solute band. The fraction of solute originally sorbed in plate 0 and which is now in solution in the Pth plate from the top of the column is given by

$$M_{N,P} = \frac{(N + P - 1)!}{P!\,(N - 1)!} \cdot \frac{D^{N-1}}{(D + 1)N + P} \tag{2-13}$$

where N is the number of multiples of Δv (the volume of solution in one theoretical plate) that have equilibrated with the Pth plate up to this moment.

In a very significant outcome of the plate theory, it can be shown that $M_{N,P}$ in Eq. 2-13 has its maximum when the number of plate volumes that have passed through the column is equal to the distribution ratio times the number of plates in a column or when

$$N_{\max} = PD \tag{2-14}$$

Thus, at this point, Eq. 2-13 may be simplified (4, 6, 11, 14) so that the fraction of solute in solution in one plate volume corresponding to the zone maximum is given by

$$M_{N,P} = M_{s(\max)} = [2\pi PD(1 + D)]^{-1/2} \tag{2-15}$$

In addition to the volume represented by PD, the amount of eluent corresponding to the peak maximum may also be given as previously defined in our notations, by the expression

$$F_{\max} = \frac{N_{\max}}{P} \tag{2-16}$$

Consequently, in view of Eq. 2-14 it is readily seen that

$$F_{\max} = D \tag{2-17}$$

which means that the distribution ratio of a solute is numerically equal to the number of column volumes (number of V_0's) that have passed through the column when the concentration of a solute in the eluate is at a maximum. Equation 2-17 is a very important equation in ion-exchange chromatography

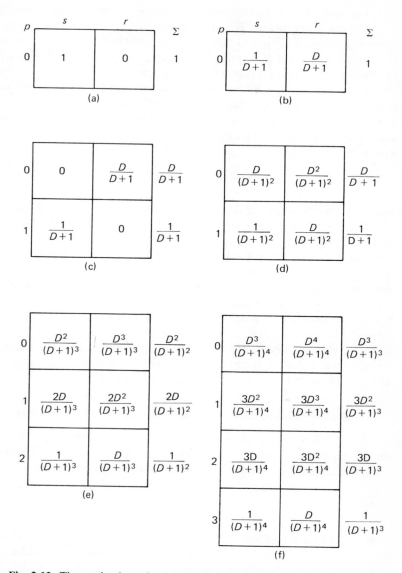

Fig. 2-13. The mechanism of solute distribution between two phases: (a) initial state before equilibration; (b) after equilibration; (c) after one transfer of solution phase, before equilibration; (d) after equilibration; (e) after two transfers of solution phase, after equilibration; (f) after three transfers of solution phase, after equilibration. See text for explanation of symbols. (Reproduced from C. J. O. R. Morris and P. Morris, *Separation Methods in Biochemistry*, Interscience Publishers, Inc., New York, 1963, p. 52, by permission of the authors and Interscience Publishers, Inc.)

45

and it is equivalent to the expressions derived for D_v by Kraus and Nelson (13) that are given in Eqs. 2-8 through 2-10, except that column volumes in those equations are given in terms of geometric column volumes instead of the column volume of Mayer and Tompkins (7), which is based upon the void volume of the column. Another difference is that, in the Mayer-Tompkins treatment, the initial void volume of the column is not included in their calculation to give the number of column volumes to reach the elution maximum. If these considerations are taken into account, then the equivalent expression for calculating the number of column volumes to the elution maximum is given by

$$F_{max} = \frac{\bar{v}}{V_0} = D + 1 \tag{2-18}$$

where $\bar{v}$ is the volume to peak, and V_0 is the void volume of the column.

Once the peak volume and the number of plates for a given column are known, the entire elution curve of a solute can be theoretically calculated not only for this column but also for other columns of different dimensions. This can be done by means of Eq. 2-13, since a plot of M_s for the last plate of the column against N is a theoretical elution curve. However, Eq. 2-13 is in a cumbersome form for calculation; also, M_s, the fraction of total solute present in one plate volume, must be transformed to give an experimentally measurable concentration, since the plate volume is in general unknown (11). Since the elution curve is bell-shaped and approximates the normal curve of error, it is possible to normalize the Mayer-Tompkins equation (given here as Eq. 2-13) to the form of a Gaussian distribution.

3. Representation of the Elution Curve as a Gaussian Error Function

This is done by simply equating the maximum ordinate at the peak of an elution curve to the peak ordinate of the Gaussian error function. The error function can be written (11) in these two equivalent forms

$$y = \bar{y} \cdot e^{-(t^2/2)} = \frac{1}{\sigma \sqrt{(2\pi)}} \cdot e^{-[(\bar{x}-x)^2/2\sigma^2]} \tag{2-19}$$

where x and y are abscissa and ordinate, respectively, and $\bar{x}$ is the abscissa of the maximum ordinate $\bar{y}$, and σ is defined as the number of standard deviations that x is removed from the peak (which in the error function is at the x origin) of the curve.

It can be shown that

$$M_{s(max)} = \frac{C_{max}V_0}{MP} \tag{2-20}$$

Therefore when this value for $M_{s(max)}$ is substituted into Eq. 2-15, the maximum ordinate at the peak of an elution curve can be expressed by

$$C_{max} = \frac{M}{V_0} \cdot \sqrt{\left(\frac{P}{2\pi D(D + 1)}\right)} \qquad (2\text{-}21)$$

It is apparent that the maximum ordinate of the error function of Eq. 2-19 is obtained when $x = \bar{x}$ and under this condition the ordinate at the peak is given by

$$\bar{y} = \frac{1}{\sigma\sqrt{(2\pi)}} \qquad (2\text{-}22)$$

By equating Eq. 2-21 to Eq. 2-22 and considering a unit of quantity of solute,

$$\sigma = V_0 \cdot \sqrt{\left(\frac{D(D + 1)}{P}\right)} \qquad (2\text{-}23)$$

By substituting for V_0 from Eq. 2-18,

$$\sigma = \bar{v} \cdot \sqrt{\left(\frac{D}{P(D + 1)}\right)} \qquad (2\text{-}24)$$

Therefore if C is the concentration of solute related to any given effluent volume v, and we let the abscissa x of Eq. 2-19 be written as $\bar{v} - v$ and, in turn, substitute the value of σ from Eq. 2-24, and if C_{max} is the maximum ordinate when $v = \bar{v}$, then the Gaussian error function of Eq. 2-19 becomes an exponential expression for a theoretical elution curve, which is represented by

$$C = C_{max}e^{-[(\bar{v}-v)/\bar{v}]^2 \cdot (P/2) \cdot [(D+1)/D]} \qquad (2\text{-}25)$$

The elution curve is now in the form of a continuous function and its entire shape can be closely approximated (when P is at least greater than 25) for a column of any given dimension, once the number of theoretical plates P is determined. The number of plates for any column for a given set of conditions can be calculated from a single exploratory elution curve (see following section). The concentration of the solute at the peak, C_{max}, and the number of column volumes to the peak, F_{max}, are also read from the elution chromatogram; once F_{max} is known, D is obtainable from Eq. 2-18. Using these values, Eq. 2-25 can be used to calculate the theoretical elution curve.

4. Methods of Determining the Number of Plates in a Column

In reference to Fig. 2-14, there are several ways to calculate P from the data of one experimental elution curve. The first of these is known as the *area-maximum ordinate* method; it involves determining C_{max} and the total

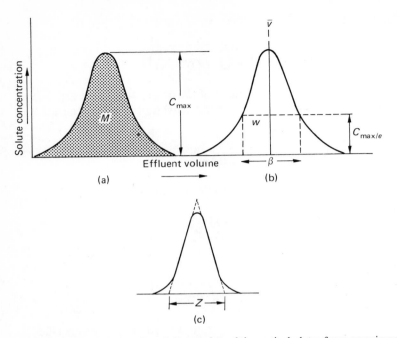

Fig. 2-14. The measurement of the number of theoretical plates from experimental elution curves. See text for proper equations that utilize the data of (a), (b) or (c) to calculate the number of theoretical plates in a column. (Reproduced from C. J. O. R. Morris and P. Morris, *Separation Methods in Biochemistry*, Interscience Publishers, Inc., New York, 1963, p. 63, slightly modified, by permission of the authors and Interscience Publishers, Inc.)

amount of solute,* experimentally, from an elution curve [see Fig. 2-14(a)] and then calculating P by means of the expression (11, 18)

$$P = 2\pi \left(\frac{C_{max} \cdot \bar{v}}{M} \right)^2 \tag{2-26}$$

The second way of calculating P is known as the *zone-width* method. By this procedure [see Fig. 2-14(b)], the half-width W of an elution curve at the ordinate value C_{max}/e is used to calculate the number of plates in a column (11). Thus

$$P = \frac{2D}{D+1} \cdot \left(\frac{\bar{v}}{w} \right)^2 \tag{2-27}$$

*Procedures for determining the amount of solute from the area of its elution curve are given in Chapter 5.

The distance W would vary if the elution curve was not exactly symmetrical; therefore, a more representative value of P is obtained if the total width β at the ordinate C_{max}/e [see Fig. 2-14(b)] is used to calculate the plate number for a column. In this case P is computed from (3, 8, 11)

$$P = 8\left(\frac{\bar{v}}{\beta}\right)^2 \tag{2-28}$$

The third method of calculating P (11, 18) as seen from Fig. 2-14(c) is the *tangent method*. The distance Z where the tangents intersect the abscissa is used for the calculation of P according to

$$P = 16\left(\frac{\bar{v}}{Z}\right)^2 \tag{2-29}$$

The plate theory predicts that P is directly proportional to the length of a packed column and is independent of its diameter. Thus, if the height of a column is held constant and its cross-sectional area is doubled, P remains constant but $\bar{v}$ for a given system is also doubled. On the other hand, if the area is held constant and the length is doubled, both $\bar{v}$ and P for the column will increase by a factor of two. Thus P and $\bar{v}$ can be calculated for columns of other dimensions once they have been determined from the elution data of a test column. However, the width and height of an elution curve varies with changes in column geometries; therefore, knowing values of P and $\bar{v}$ is not sufficient data in itself to allow one to predict the degree of resolution of a pair of solutes. For instance, given D_a and D_b for the solutes a and b, an estimate of the separation of their respective peaks can be expressed by (19, 20)

$$\bar{v}_a - \bar{v}_b = V_0(D_a - D_b) \tag{2-30}$$

where V_0 is the void volume of a column. When $\Delta D_{ab} = (D_a - D_b) = 0.1$, peak separation is $0.1V$ or $\frac{1}{10}$ of a column volume;* this gives an estimate of the minimum column volume necessary for a reasonable separation. Nevertheless, from this limited data, the degree of resolution cannot be ascertained since the width of the elution curves remains unknown. However, by calculating the entire elution curves for a pair of solutes, the widths of the curves and hence the degree of overlap can be predicted.† In this respect, upon evalu-

---

*If the number of column values are expressed in geometric column volumes, V_0 of Eq. 2-30 is replaced by the term area $(A) \cdot$ length (l) and the volume distribution coefficient D_v of Kraus and Nelson is used instead of the Mayer-Tompkins coefficient D. These relationships can be seen from Eqs. 2-7 and 2-10 and Eqs. 2-17 and 2-18.

†Together with knowing their distribution coefficients, once the respective widths of two elution curves for a pair of solutes is known, separation data for the solutes can be expressed in terms of a resolution factor; such correlations are given in Chapter 5.

ation of the data from a test column, other values of $\bar{v}$ and P may want to be considered. For these newer values, C_{max} may be calculated from

$$C_{max} = \frac{M}{\bar{v}} \cdot \left(\frac{P}{2\pi}\right)^{1/2} \qquad (2\text{-}31)$$

Thus, Eq. 2-25 may be applied to calculate elution curves under the conditions of these newer parameters.

5. The Concept of the Plate Theory as a Continuous Process

Applications of the plate theory have been increased in value due to refinements by Glueckauf (8). The main objection to the Mayer-Tompkins approach in developing their plate theory is that they visualized flow through the column as a series of discontinuous movements in which the free finite volume of a plate is equilibrated step by step with an amount of resin in each successive plate of finite length. This approach is inherent in the mathematical treatment given the plate theory when it was developed with the aid of Fig. 2-13 and Eq. 2-13. In reality, even though the resin beads are discontinuous, liquid flow through the column is a continuous process. On this basis, Glueckauf developed a newer theoretical plate model.

The initial premise of both models is the same in that the column is considered as being divided into units of length called plates in which the concentration of solute is uniform both in the resin phase and the solution phase and that the two concentrations are assumed to be in equilibrium. Glueckauf measures the length of a plate and the distance from the top of the column in units of a geometric column volume [i.e., area $(A)\cdot$length (L) = column volume (X)] so that distance can be measured in terms of x, which is a fractional value of X, or of l, which is a fractional value of L. Thus, if Δl is the thickness of a theoretical plate and the column is A cm^2 in cross-sectional area, the volume of a plate will be $\Delta x = A\ \Delta l$.

The mathematical approach begins as follows (8, 14). Let x be that volume of column above a theoretical plate with the very small volume Δx. Next imagine that a volume v has already passed through the column and therefore has gone through the plate. If C_x is the concentration of solute in the plate and q_v is the total concentration in both the resin and solution phases and an additional small volume of solution Δv is then allowed to flow through the plate, it follows that the net amount of solute transferred to the plate is equal to the difference between the amount entering and leaving the plate or

$$\Delta v[c_{(x-\Delta x)} - c_x] = \Delta x[q_{(v+\Delta v)} - q_v] \qquad (2\text{-}32)$$

Equation 2-32 is the foundation upon which the continuous flow model is considerably expanded, mathematically, by methods of differential calculus; only the resultant equations pertinent to practical application of the plate theory will be presented here.

A simple relationship that arises from the mathematics given to the continuous flow model is that the number of column volumes required for a solute to reach its maximum concentration in the effluent is equal to a constant or

$$\frac{\bar{v}}{X} = a \tag{2-33}$$

where a is numerically equal to a distribution ratio f/c, whose components are defined as follows:

$f =$ amount of solute in both phases per cm^3 of column volume

$c =$ concentration of solute in aqueous phase

From equations previously given, a is related to the volume distribution coefficient D_v by the following:

$$a = \frac{f}{c} = \frac{\bar{v}}{X} = D_v + \epsilon \tag{2-34}$$

Equations 2-33 and 2-34 are equivalent to the Mayer-Tompkins relations given in Eqs. 2-17 and 2-18, wherein the number of column volumes necessary to reach the elution maximum is equal to a distribution ratio. Also as in the discontinuous model for the plate theory, equations from the newer refinements due to Glueckauf can be normalized in terms of the normal curve of error so that the amounts of solute in an elution zone and the shape of the elution curve can be conveniently computed. The exponential form for an elution curve that results from the mathematical treatment given to the continuous model of the plate theory is expressed by

$$C = C_{max}e^{-[(\bar{v}-v)^2/(\bar{v}\cdot v)]\cdot(P/2)} \tag{2-35}$$

Since the exponential terms of both Eq. 2-25 and Eq. 2-35 approach zero when v approaches the value of $\bar{v}$, the values for the peak position and the maximum concentration are the same from either equation. At places other than near the elution maximum the curves of Eq. 2-25 and Eq. 2-35 are not identical, even when D is larger, due to the difference in the exponential terms of the equations. As a result of this difference, Eq. 2-35 predicts that the leading edge of an elution curve will be somewhat sharper than the trailing boundary, while Eq. 2-25 predicts a symmetrical curve (11). In practice, the bands from the Mayer-Tompkins treatment are flatter and broader than the theory predicts (14).

Just as in the discontinuous approach to the plate theory, an estimate of

P and the determination of C_{max} must be obtained from experimental data (e.g., see the elution curves of Fig. 2-14) before the shape of an entire elution curve can be computed by means of Eq. 2-35. The number of theoretical plates may be calculated, as before, from Eq. 2-26 through Eq. 2-29.

On the assumption that two solutes are present in equal amount, Glueckauf (8) also derived a simple expression to determine the elution volume at which the impurity ratios for two solutes are at a minimum in the region where their elution curves overlap. Let M_1 and M_2 represent the amount of the solutes and their elution curves be those shown in Fig. 2-15. If the curves are of equal area and the cross-contamination is small enough, then

$$\frac{\Delta M_1}{M_2} = \frac{\Delta M_2}{M_1} = \eta \qquad (2\text{-}36)$$

where ΔM_1 is the amount of M_1 in the elution band of M_2 and ΔM_2 is the amount of M_2 in the elution zone M_1; the percentage purity of both solutes is then $100(1 - \eta)$. Minimum ratios of Eq. 2-36 occur when the volume to the point of intersection of the elution curves is given by

$$v_1 = \sqrt{\bar{v}_1 \cdot \bar{v}_2} \qquad (2\text{-}37)$$

As the ratio of M_1 to M_2 changes, calculations become more complicated

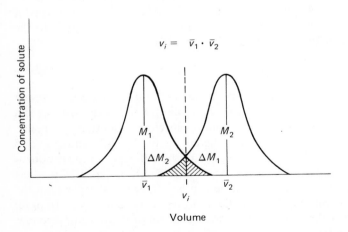

Fig. 2-15. Point of intersection where cross-contamination of overlapping elution bands is at a minimum.

and the use of Eq. 2-36 to calculate impurity ratios becomes less and less accurate. An original formula derived by Glueckauf (8) to calculate impurity ratios for the cases when M_1 is not equal to M_2 was found by Said (21) to be in error. The latter worker corrected the Glueckauf equation and he presents a method by which the efficiency of chromatographic separations can be obtained from a chart wherein the ratio of M_1 to M_2 is varied considerably.

6. Parameters That Affect the Performance of a Column

The degree of separation for two solutes will depend upon the widths of their elution curves and this, in turn, depends upon the number of exchanges or number of plates encountered by a solute ion as it passes down a column. Accurate predictions of solute behavior by the plate theory are only possible when the spreading of elution zones is *not* caused by lack of an equilibrium within the column. For a resin bed of any size, the number of plates in the column will depend upon the operating conditions, and any condition that tends to promote equilibrium will decrease the HETP and hence increase the fractionating efficiency of the column, i.e., increase the number of plates per unit length of a column. It is generally assumed that the effective plate height is the same for different ionic species, but this is not strictly true, since spatial dimensions and other properties of individual solutes, such as charge, will probably influence the rate at which equilibrium is attained (11, 22).

As a solute moves through an exchanger bed, the number of theoretical plates it encounters is determined by the following factors: (1) column length; (2) flow rate of eluent; (3) particle size of the exchanger; (4) temperature.

The number of plates increases with the height of an exchanger bed. However, in practice the length of a column that can be used is limited by the fact that in addition to diffusion and thermal mixing, the plate theory predicts that the height of an elution curve decreases inversely with the square root of column length while its width increases directly with the square root of column length (3).

Fast flow rates decrease the degree of separation for a pair of solutes, since this retards establishment of equilibrium between the solution and resin phases of the column and in effect reduces the number of plates in the exchanger bed. For slower flow rates that enhance the condition of equilibrium, the height of a theoretical plate is approximately equal to the diameter of the resin beads in a column (23).

Decreasing the particle size of the exchanger beads reduces the distance an ion must diffuse in both the aqueous and resin phases to effect an exchange or transfer; this promotes the rate at which equilibrium is attained between the two phases.

An increase in temperature increases the rate of diffusion as well as the rate of reaction in a column. By operating at higher temperatures, the number of plates may be increased several fold over the number it has when operated at room temperature. Therefore a shorter column can be employed at higher temperature to achieve a given separation that would require a much longer column, and would, perhaps, result in poorer resolution, if the separation were carried out at room temperature.

In most separation work, the experimental elution curve can be made to approach the shape of the theoretical elution curve. The two curves coincide only when the experimental circumstances conform to the following set of conditions: (1) if there is uniform packing within the column; (2) if low solute concentrations are employed; (3) and if the column system is optimal with regard to the other operational factors just discussed, such as column length, flow rate, particle size of the exchanger, and temperature.

REFERENCES

1. Kunin, R., *Ion Exchange Resins*, John Wiley & Sons, Inc., New York, 1958, 2nd ed., Chap. 6, p. 115.

2. "Recommendations on Ion-Exchange Nomenclature," Pure and Applied Chem., **29**, 619 (1972).

3. Samuelson, O., *Ion Exchange Separations in Analytical Chemistry*, John Wiley & Sons, Inc., New York, 1963, Chaps. 5, 6.

4. Berg, E. W., *Physical and Chemical Methods of Separation*, McGraw-Hill Book Company, Inc., New York, 1963, Chap. 10.

5. Schubert, J., in *Methods of Biochemical Analysis*, Interscience Publishers, Inc., New York, 1956 (ed. by D. Glick), Vol. 3, p. 247.

6. Tompkins, E. R., and Mayer, S. W., *J. Am. Chem. Soc.*, **69**, 2859 (1947).

7. Mayer, S. W., and Tompkins, E. R., *J. Am. Chem. Soc.*, **69**, 2866 (1947).

8. Glueckauf, E., *Trans. Faraday Soc.*, **51**, 34 (1955).

9. Tompkins, E. R., *J. Chem. Education*, **26**, 32 (1949).

10. Reichenberg, D., in *Ion Exchangers in Organic and Biochemistry*, Interscience Publishers, Inc., New York, 1957 (ed. by C. Calmon and T. R. E. Kressman), Chap. 4.

11. Morris, C. J. O. R., and Morris, P., *Separation Methods in Biochemistry*, Interscience Publishers, Inc., New York, 1963, Chap. 2.

12. Rieman III, W., and Sargent, R., in *Physical Methods in Chemical Analysis*, Academic Press, Inc., New York, 1961 (ed. by W. G. Berl), Chap. 5.

13. Kraus, K. A., and Nelson, F., *Symposium on Ion Exchange and Chromatography in Analytical Chemistry*, Am. Soc. Testing Mater., Spec Publ. No. 195 (June 1956), p. 27.

14. Walton, H., in *Chromatography*, Reinhold Publishing Corporation, New York, 1967 (ed. by E. Heftmann), 2nd ed., Chap. 12.

15. Cornish, F. W., *Analyst*, **83**, 634 (1958).

16. Rieman III, W., *Record Chem. Progr.* (Kresge-Hooker Sci. Lib.), **15**, 85 (1954).

17. Martin, A. J. P., and Synge, R. L. M., *Biochem. J.*, **35**, 1358 (1941).

18. Cohn, W. E., *J. Am. Chem. Soc.*, **72**, 1471 (1950).

19. Hamilton, P. B., Bogue, D. C., and Anderson, R. A., *Anal. Chem.*, **32**, 1785 (1960).

20. Hamilton, P. B., in *Advances in Chromatography*, Marcel Dekker, Inc., New York, 1966 (ed. by J. C. Giddings and R. A. Keller), Vol. 2, Chap. 1.

21. Said, A. S., *J. Gas Chromatog.*, **1**, 20 (1963).

22. Helfferich, F., *Ion Exchange*, McGraw-Hill Book Company, Inc., New York, 1962, Chap. 9.

23. Glueckauf, E., *Ion Exchange and Its Applications*, Society of Chemical Industry, London, 1955, p. 34.

Selecting the Proper
Ion-Exchange Material
THREE

A. PRELIMINARY CONSIDERATIONS

1. Limitation of Ion-Exchange Materials to Four Classes

In addition to the synthetic resins described in Chapter 1, i.e., those prepared by vinyl polymerization, there is a wide variety of other ion-exchange materials commercially available. In this book we limit these other materials to the inorganic ion exchangers, the ion-exchange celluloses, and the ion-exchange dextrans and polyacrylamide gels. Except for very specialized cases, these three classes of ion-exchange materials, together with the vinyl-crosslinked resins, readily meet most requirements of ion-exchange work. Even with these limitations, the problem of choosing a suitable exchanger for a particular purpose, at first glance, seems like a formidable task. For instance, the user of ion-exchange materials, in comparing the exchangers of each class, has to evaluate such properties as particle size, porosity, type of functional group attached to a matrix, the chemical nature of the matrix material itself, exchange capacity, and thermal and chemical stability. However, when these very properties are correlated with the chemical and physical properties of the sample to be processed by ion exchange, the selection of a proper ion-exchange material is narrowed to only a few choices.

56

2. Selecting the Exchanger by a Process of Elimination

The selection of an exchanger is simple if a procedure taken from the literature is to be followed, since the trade name and type of exchanger is usually given in the original publication. If the identical exchanger is no longer available, it is then necessary to purchase a newer material or to buy an equivalent exchanger as sold by another vendor. Later in this chapter, where each class of exchange material is discussed individually, the sources as well as the properties of commercially available exchangers are tabulated; equivalent products (or nearly equivalent products) as sold by different vendors can be compared.

When it is necessary to extend other work or to establish new ion-exchange methods, a simple process of elimination dictates what type of exchange material will be needed. The elimination process consists of asking three fundamental questions about the properties of a sample under investigation: what is the net charge of the solutes under study? what is the molecular weight of the solutes? what is the chemical and physical environment of the solvent system?

(a) *Selection based on the net charge of a solute* Solely on the basis of function (i.e., the basic mode of operation of a cation or anion exchanger), the answer to the first question of the elimination process excludes half of the ion-exchange substances available for ion-exchange work. This follows because by knowing whether the solutes of interest are cationic or anionic at a given pH, the user of ion-exchange materials chooses an exchanger in accordance with the following chart.

Ion exchanger

Cation exchanger			Anion exchanger		
Strong acid	Intermediate acid	Weak acid	Strong base	Intermediate base	Weak base

Only when the isoelectric point of a molecule lies near neutrality and a stable ion exists on either side of this point does one have a choice between a cation or anion exchanger. The choice of acid or base strength that concerns the selection of either a cation or anion exchanger rests mainly on the pH's that can be tolerated during the ion-exchange procedure. Strong-acid or strong-base exchangers show ion-exchange behavior over the full pH range. However, a weakly acidic cation exchanger in the H-form does not react with the cation of a neutral salt; it has a very high affinity for hydrogen ions and its exchange capacity decreases rapidly below pH 4. Similar properties are associated with weakly basic anion exchangers, they are highly ionized only in a salt form and have little exchange activity above pH 7. Selectivities and capacities also vary in different acid- or base-strength exchangers. This information is usually compared in brochures published by suppliers of ion-exchange materials.

(*b*) *Selection based on the size and net charge of a solute*   Consideration of molecular weight (the size of a molecule) as well as net charge is a very important factor in the elimination process. As the size and net charge of molecules increase, the general usefulness of the synthetic resins (the vinyl-crosslinked types) becomes more and more limiting. As seen in Chapter 1, these exchangers are visualized as crosslinked polyelectrolytes whose exchange groups become hydrated when the resins are placed in water. These ion-exchange materials would imbibe water and swell indefinitely and eventually become soluble if it were not for the covalent vinyl crosslinks holding the polymeric polyelectrolyte chains together (1, 2). Hence, the elastic forces of the polymer framework restrict the porosity of these exchangers so that their pore size varies from a few angstrom units (Å) for the highly crosslinked resins up to only about 50 Å for the weakly crosslinked polymers (3). Due to such restrictions, a point is reached where increasingly larger molecules cannot move freely in and out of the criss-crossing polymer network of a resin; this *ionic sieve effect* is shown schematically in Fig. 3-1. As a consequence of this screening effect, the probability of an ion entering a pore is small unless the pores of the exchanger are at least three times the diameter of the molecule (4). Therefore for a resin to function in the manner intended, it is apparent that the degree of crosslinking must be chosen so that the ions under study are able to diffuse freely in and out of the interior of a resin particle. This automatically prohibits, on the basis of size only, the use of many of the vinyl-crosslinked exchangers for the study of the larger organic solutes.

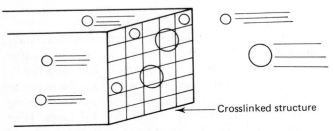

Crosslinked structure

Fig. 3-1. Screening (ionic sieve effect) of large ions by crosslinked ion-exchange structure. (Reproduced from R. Kunin, *Elements of Ion Exchange*, Reinhold Publishing Corporation, New York, 1960, p. 18, by permission of the author and the Reinhold Publishing Corporation.)

An additional factor comes into play if the total net charge of large molecules also increases as they become larger in size. Such polyvalent species as peptides and polynucleotides (depending on their size) can become irreversibly bound to the resinous exchangers. Since these types of organic solutes and the resin exchangers that attract them are both considered as polyelectrolytes—the fixed ionic groups of the exchanger being opposite in charge to those contained in the polyvalent organic ions—a large number of ionic bonds are involved in the mutual forces of attraction. Consequently, a large number of bonds have to be broken simultaneously by replacing ions in order for desorption to take place. In such circumstances, the position of equilibrium is such that the polyvalent solute favors the exchanger phase, not the solution phase; hence, in these cases irreversible sorption occurs. As a consequence of the ionic sieve effect or of the high probability of multiple bonding, or of both of these occurring, not many organic solutes with a molecular weight of over 500 can undergo "true" ion-exchange reactions with the vinyl type exchangers (2, 4). However, as seen in Chapters 8 and 9, the resinous exchangers may be utilized as the sorbent material for chromatographic processes that take place by other mechanisms such as surface sorption, partition, ion exclusion, ligand exchange, etc.

For organic molecules of molecular weight greater than 500 on up to such macromolecules as serum proteins, nucleic acids, and enzymes, chromatography and other analytical operations are customarily carried out on the ion-exchange celluloses, dextrans, or similar exchangers. Compared to the vinyl type exchangers, these materials have a greater distance between the incorporated exchange sites on their polymeric matrices. As a result of this lower density of binding sites (i.e., lower exchange capacity per unit weight of material), the probability of multiple binding occurring between large molecules and the exchangers is correspondingly lessened and in most cases, even with the macromolecules, reversible equilibria can be established. Furthermore, the processing of high molecular-weight polyvalent substances by ion-exchange methods usually requires either that the majority of the fixed ionic groups appear on the outside of the exchanger's framework or that the interior of the exchanger is readily accessible to the large solutes. The ion-exchange celluloses are representatives of the first mentioned category and the ion-exchange dextrans and polyacrylamide gels are examples of crosslinked substances that have pores of such size that even macromolecules can penetrate into the interior of these gels.

The large pore size, the enormous surface area, the low number of exchange sites and their hydrophilic nature are the favorable characteristics that make the polyacrylamide and dextran ion-exchange gels, together with the cellulose exchangers, superior materials for the processing of macromolecules. In such favorable surroundings, macromolecules have little tendency to become denatured. The low capacity of these exchangers lessens the probability that their exchange sites will compete with the multitude of weak bonds that maintain the secondary structure of a macromolecule, and there are no lipophilic surfaces to disorient the native configuration of a large polymeric electrolyte (4, 5).

(c) *Selection based on the chemical and physical environment of a solute* In some circumstances, only one class of exchanger is applicable to the problem at hand. All other ion-exchange substances except *the inorganic ion exchangers* would be eliminated as possible exchange materials for the treatment of a group of ions if the ions were contained in any of the following environments: (1) high levels of radiation; (2) temperatures maintained at over 100°C; (3) concentrated solutions of potent oxidizing or reducing agents; (4) hot concentrated alkaline or acidic solutions.

Exchangers that can withstand such drastic surroundings are synthetic microcrystalline inorganic aggregates prepared by the careful precipitation of certain hydrous oxides, acid salts, or salts of heteropolyacids (6, 7, 8). Precipitated hydrous oxides of tetravalent metals such as zirconium, thorium, titanium, and tin behave as cation exchangers in basic solutions and as anion exchangers in acidic solutions. A different set of cation exchangers, with excellent properties, may be prepared by combining the acid oxides of tetravalent metals (e.g., ZrO_2^{++}) with polyvalent anions such as phosphate, molybdate, tungstate, or arsenate. Also used as cation exchangers are the insoluble salts of heteropolyacids such as microcrystalline ammonium phosphomolybdate or phosphotungstate. All these crystalline ion exchangers have repeating units that comprise a rigid polymeric framework. This is in contrast to the swelling and shrinking of the organic type exchangers with their flexible matrices.

Because of their inorganic composition, the microcrystalline ion-exchange gels can function in extremely harsh environments that would destroy any organic ion-exchange material, if present, in a few minutes. For instance, the ion-exchange crystals are capable of performing separations at temperatures up to 300°C and are able to withstand the high levels of radiation found in reactor wastes (6, 7, 8, 9). Another unique property of these exchangers is their ability to exhibit extremely high affinities towards certain ions such as cesium, rubidium, and strontium; also, inorganic ion exchangers are able to exchange effectively in nonaqueous solutions (6, 9).

B. THE ION-EXCHANGE RESINS

1. Trade Names, Manufacturers, and Suppliers

In the literature, ion-exchange resins are identified more often by trade names than by chemical terminology. For this reason and for the sake of convenience in referring to them in other instances, not only their chemical constitution but also the various trade names for equivalent or identical exchangers should be known.

Since exchangers with monofunctional groups are preferred for both chromatographic and batch procedures, it follows that, for all practical purposes, only four types of resinous exchangers need to be considered in detail. All four contain divinylbenzene as the crosslinking agent. Three have styrene-divinylbenzene matrices to which are bonded, respectively, sulfonic acid, quaternary ammonium, and tertiary amine groups, and they are classified in the order given as a strongly acidic cation exchanger, a strongly basic anion exchanger, and a weakly basic anion exchanger. The other exchanger has carboxylic acid groups attached to an acrylic-divinylbenzene network and is classified as a weakly acidic exchanger (chemical structures for these exchangers are given in Chapter 1). Some of the more widely known manufacturers of these ion-exchange resins and the corresponding trade names for them are given in Table 3-1; more complete lists can be found elsewhere (e.g., see Chapter 9 of reference 4, or the appendices of references 10 and 11). Some less commonly used resins are described together with a few miscellaneous ion-exchange substances in a later section of this chapter.

If purchased directly from the manufacturer, it is most likely that a resin will have to be pretreated before it is put to use. This, in the main, consists of separating (usually by hydraulic means) the very fine particles from the very large ones, and of washing the exchanger, consecutively, with different solvents in order to remove any reaction products that may have been trapped in the beads during manufacture. These steps are time consuming and for general laboratory use it is often more expedient to purchase resins that have been pretreated. At added cost, certain suppliers offer a variety of ion-exchange resins that have been purified* and that are available in specified narrow ranges of particle size. In the United States most of the resins listed in Table 3-1 are obtainable, in a refined condition, from the Mallinckrodt Chemical Company (St. Louis, Mo.), from the Bio-Rad Corporation (Richmond, Calif.), and from more general chemical supply companies such as J. T. Baker or Fisher.

2. Terminology

Since there are so many variations among each kind of exchanger, complete identification cannot be made solely on the basis of a trade name or chemical name. It is also necessary to know the particle size, the degree of crosslinkage, the capacity, and the ionic form of a resin before it can be fully characterized.

---

*Even resins that have been purified, when not used for a while, often give rise to colored impurities; the leaching of such impurities, known as "color throw," subsides each time the resins are rinsed anew and generally ceases after a few column runs have been made.

TABLE 3-1
Trade Names and Manufacturers of the Most Common Commercially Available Ion-Exchange Resins[a]

Resin type	Chemical constitution	Usual form as purchased	Trade names[b] of equivalent ion exchangers					
Strongly acidic cation exchanger	Sulfonic acid groups attached to a styrene and divinylbenzene copolymer	ϕ—$SO_3^-H^+$	Amberlite IR-120	Dowex 50W	Duolite C-20	Lewatit S-100	Ionac C-240 (or Permutit Q)	Zeocarb 225
Weakly acidic cation exchanger	Carboxylic acid groups attached to an acrylic and divinylbenzene copolymer	R—COO^-Na^+	Amberlite IRC-50	—	Duolite CC-3	Lewatit C	Ionac C-270 (or Permutit Q-210)	Zeocarb 226
Strongly basic anion exchanger	Quaternary ammonium groups attached to a styrene and divinylbenzene copolymer	[ϕ—$CH_2N(CH_3)_3$]$^+Cl^-$	Amberlite IRA-400	Dowex 1	Duolite A-101D	Lewatit M-500	Ionac A-450 (or Permutit S-1)	Zeocarb FF (or De-Acidite FF)
Weakly basic anion exchanger	Polyalkylamine groups attached to a styrene and divinylbenzene copolymer	[ϕ—$NH(R)_2$]$^+Cl^-$	Amberlite IR-45	Dowex 3	Duolite A-7	Lewatit MP-60	Ionac A-315 (or Permutit W)	Zeocarb G

[a]Compiled from tables found in references 9, 10, and 11.
[b]Manufacturers: Amberlite, Rohm & Haas Co., Philadelphia, Pa., U.S.A.; Dowex, Dow Chemical Co., Midland, Mich., U.S.A.; Duolite, Diamond Alkali Co., Redwood City, Calif., U.S.A.; Lewatit, Farbenfabriken Bayer, Leverkusen, Germany; Ionac (or Permutit of U.S.A.), Ionac Chemical Co., New York, N.Y., U.S.A.; Zeocarb, The Permutit Co. Ltd., London, England.

(a) Particle size Resin particles or beads can be formed with diameters ranging from below 1 μ (colloidal) up to several millimeters in size (10). However, for most applications resins are chosen that have particle sizes that lie in between these extremes (11). The particle sizes are usually given in terms of a mesh size specified by a range of U.S. standard screens. Mesh designations as well as other units for expressing particle size are shown in Table 3-2.

TABLE 3-2

Some Common Ranges of Particle Size for Resins

Mesh range Designation	Screen analysis	Diameter of particles		
		Inches	mm	Microns
16– 20	Wet	0.0460–0.0331	1.168–0.84	1168–840
20– 50	Wet	0.0331–0.0117	0.84 –0.297	840–297
50–100	Dry	0.0117–0.0059	0.297–0.149	297–149
100–200	Dry	0.0059–0.0029	0.149–0.074	149–74
200–400	Dry	0.0029–0.0015	0.074–0.038	74–38

Reproduced from *Dowex:: Ion Exchange*, © 1958, 1959, 1964, by permission of the Dow Chemical Company.

For very small diameter particles the mesh size is based upon a wet screen analysis of the final product, while for the larger sized resin beads the mesh designation is for the dry polymers before functional groups have been introduced. In the latter case the beads, when immersed in water, are larger than indicated by the container label.

As the size of the resin particle is decreased, the following effects are noted: the time required to approach equilibrium with a contacting solution is decreased; the pressure differential through a column increases, hence the flow rate decreases; the number of "theoretical plates" in a column of constant height increases, or conversely, the volume of resin required for a specific operation decreases; and the settling rate of the resin decreases.

As a consequence of these effects, it has been found that 100–200 mesh particles are of suitable size for batch work and for general column operations, but that at least 200–400 mesh resin particles should be used for separation work in low pressure (gravity feed) chromatographic columns. Very coarse beads, < 100 mesh, are reserved for large-scale operations where fast settling rates or fast flow rates are required. Only the very fine particles, 1–20 μ (colloidal range) are utilized in high pressure, high resolution chromatographic apparatus. Such systems are expensive, consist of intricate modules, and are designed primarily for the analysis of complicated biochemical or clinical samples. However, some of the ultimate goals of chromatography are achieved by these high performance systems and they are further discussed in Chapters 4 and 7.

(b) *Per cent crosslinkage and capacity*   The per cent divinylbenzene (% DVB) in an exchanger is expressed by the symbol -X followed by a number. The trade name for a particular exchanger usually precedes the "-X" number; thus, Amberlite IR-120-X8 and Dowex 1-X12 are expressions that describe resins made from copolymers containing 8% and 12% DVB, respectively.

The fraction of DVB in a resin determines to what extent the exchanger is free to expand or contract. This, in turn, is what determines the "pore" size of the matrix. Resins with a lower DVB content are able to swell more than those with a higher percentage of DVB, hence low crosslinked resins are more porous.

As the crosslinkage increases, the diffusion of ions within the beads becomes slower, the time required to approach equilibrium with a surrounding solution is slower, and the exchanger increases its tendency to keep large ions from entering its interior. Conversely, as the crosslinkage is decreased, these properties are reversed.

Because of the inverse relationship between crosslinkage and swelling, the DVB content of the styrene type strongly acidic and strongly basic exchangers affects the capacity of these resins in two ways (11, 12, 13). As the degree of crosslinkage increases, a resin will have a smaller swollen volume, for essentially the same number of exchange sites, than it had at a lower DVB content. Consequently, the *wet-volume capacity* (meq per ml of wet resin) increases. An opposite effect, but minor in comparison, is noted for the *dry-weight capacity* (meq per g of dry resin) of an exchanger. As the degree of crosslinkage increases, the dry-weight capacity decreases, since due to steric effects there is a greater difficulty of substituting functional groups in the aromatic rings of the styrene-divinylbenzene matrix (12). A typical example of how the capacity varies as the % DVB is changed can be seen from the capacity ratings of a sulfonated polystyrene cation exchanger. As the crosslinkage is increased from -X1 to -X16, the wet-volume capacity for the H-form of this exchanger increases from 0.4 to 2.4 meq per ml of wet resin, while the dry-weight capacity decreases only from 5.4 to 5.1 meq per g of dry resin (13). Methods for determining the total ion-exchange capacity of exchangers are given in Appendix B.

(c) *The ionic form of an exchanger*   The chemical identity of the counterion, i.e., the ionic form of an exchanger, must be known before a resin can be properly utilized. However, the ionic form of an exchanger is one of the least important marks of identification, since an exchanger can be completely converted from an unknown ionic form to a known form by simply flowing the proper salt solution through it.

Cation exchangers are usually sold only in their H- or Na-forms and anion exchangers in their Cl-forms. However, for a service fee, some suppliers of ion-exchange materials will convert resins to any form that the customer requests.

(*d*) *Formulary* The notations of ion-exchange terminology are ordinarily used in combination so that the essential characteristics of a resin can be indicated with only a few short expressions and symbols. Thus, from the nomenclature of I, II, or III,

Amberlite IRA-400-X8
200–400 mesh
Cl-form
$[RN(CH_3)_3]^+Cl^-$

I

Duolite C-20-X10
100–200 mesh
Na-form
$\phi SO_3^-Na^+$

II

Dowex 3-X4
50–100 mesh
Cl-form
$[\int NH(CH_3)_2]^+Cl^-$

III

the trade name, per cent crosslinkage, the particle size, the ionic form, and the type of the exchanger are all readily indicated. The symbols R, ϕ, $\int$, or other convenient characters, are used to indicate the matrix material of an exchanger; the choice is purely arbitrary. Often, since the type of exchanger is implicit by its trade name, the formula of a resin will only contain a symbol for the matrix material and the chemical abbreviation for the counter-ion, e.g., I and II could be written simply as RCl and ϕNa, respectively.

3. Chemical and Physical Stability

Chemical and thermal degradation of the matrix material are the chief causes for resin deterioration (10); stabilities vary with the resin type, temperature, ionic form, and pH. The vinyl-crosslinked resins can withstand most reducing agents and will even tolerate moderate concentrations of some oxidizing agents, e.g., nitric acid, 2–3 *M*, at room temperature (11). The hydroxide forms of the vinyl type anion exchangers are stable only up to about 30°C, whereas most of the salt forms of these anion exchangers as well as the common salt and hydrogen forms of the corresponding cation exchangers are stable at temperatures exceeding 100°C (9). Samuelson (11) discusses the analytical limitations of the resinous exchangers as used in the presence of oxidizing and reducing agents; other general thermal and chemical properties of some exchangers are given in Table 3-3.

TABLE 3-3

Chemical and Physical Properties of Ion-Exchange Resins[a]

Resin type	Matrix material	Function group	Order of selectivity for monovalent ions . . . for divalent ions	Thermal	Stability		Reduction
					[1]Solvent (alcohols, hydrocarbons, etc.) [2]Oxidation . . .		
Strongly acidic cation exchanger	Polymerized styrene-divinylbenzene	ϕ—SO_3^-	Ag > Rb > Cs > K > NH$_4$ > Na > H > Li Zn > Cu > Ni > Co	Good up 150°C	[1]Very good [2]Slow solution in hot 15% HNO_3		Very good
Weakly acidic cation exchanger	Polymerized methacrylic acid-divinylbenzene	R—COO$^-$	H ≫ Ag > K > Na > Li H ≫ Fe > Ba > Sr > Ca > Mg	Good up to 100°C	[1]Good [2]Good		Good
Strongly basic anion exchanger	Polymerized styrene-divinylbenzene	[ϕ—$CH_2N(CH_3)_3$]$^+$	I > phenolate > HSO$_4$ > ClO$_3$ > NO$_3$ > Br > CN > HSO$_3$ > NO$_2$ > Cl > HCO$_3$ > IO$_3$ > HCOO > Ac > OH > F	OH$^-$ form fair up to 50°C Cl$^-$ and other forms good up to 150°C	[1]Very good [2]Slow solution in hot 15% HNO_3 or conc. H_2O_2		Good[b]
Weakly basic anion exchanger	Polymerized styrene-divinylbenzene	[ϕ—$NH(R)_2$]$^+$	[c]ϕSO_3H > HCit > CrO$_3$ > H_2SO_4 > tartaric > oxalic > H_3PO_4 > H_3AsO_4 > HNO$_3$ > HI > HBr > HCl > HF > HCOO > HAc > H_2CO_3	Extensive information not available. Tentatively limited to 65°C	[1]Very Good [2]Good		Unknown

[a]Selectivity and stability data from reference 9 reproduced here, in part, by permission of the Bio-Rad Corp., Richmond, Calif.
[b]Will break down in the presence of sulfur-containing reducing agents.
[c]Sequence for acid solution only.

C. ION-EXCHANGE CELLULOSES

1. Fibrous Cellulosic Exchangers

Cellulose as isolated from its natural sources (e.g., from wood or cotton) has a fibrous macrostructure. Aggregates of glucosidic chains in various states of order and disorder are randomly oriented along the fiber axes (14, 15). The higher-oriented chains are described as fibrillar "crystalline" areas or bridges and these dense centers are interconnected by longer axial fibers composed of low-ordered amorphous glucosidic chains (14, 15, 16). Hydrogen bonding between the neighboring cellulose chains, and especially in the fibril centers, provides dimensional stability for the cellulose matrix, and it is these forces that restrict the matrix to only moderate swelling and make cellulose insoluble in water (14, 17).

Within and in between the cellulose chains are located "holes" or pores with a wide range as to size (14, 15). When ionizable groups are introduced into such a matrix, the natural polymer cellulose becomes an ion-exchange material. Figure 3-2 shows, diagrammatically, the microstructure of a cellulosic exchanger.

Chemical reaction to attach ionized groups to the cellulose matrix proceeds with difficulty in the crystalline regions, but takes place more readily in the amorphous areas. The substitution of functional groups into cellulose

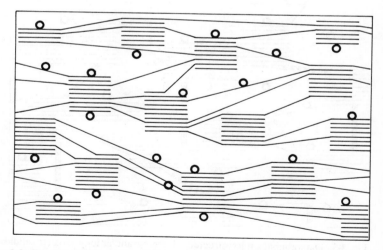

Fig. 3-2. Diagrammatic microstructure of fibrous ion-exchange celluloses. The solid lines represent the aggregates of carbohydrate chains. The dotted circles are the ion-exchange sites. (Reproduced with permission from Reeve Angel Scientifica Division, supplier of Whatman advanced ion-exchange celluloses.)

has a disruptive effect on its structure. If carried out to completion, the cellulose matrix would be destroyed and ultimately water-soluble polymers would be formed (14). Even at high levels of substitution, much before solubility occurs, cellulose derivatives lose their attractiveness as a chromatographic material for the fractionation of large solutes. A high density of uniform charges makes difficult the elution of large polyelectrolytes, or even may prevent the removal of macromolecules from highly substituted exchangers. Therefore chemical reactions to introduce functional groups are not carried to completion. Substitution is restricted to the more reactive centers (e.g., the most accessible regions) in the amorphous regions and is seldom carried out beyond the level of 1 meq per g of dry exchanger (4, 14, 17). At this level of substitution, the native configuration of the cellulose structure is only slightly modified and the low density nonuniform exchange sites are readily accessible to large polyelectrolytes.

2. Microgranular Cellulosic Exchangers

The characteristic structure of cotton cellulose can be modified by mild acid hydrolysis. Under such treatment chain splitting and recrystallization occurs within the interfibrillar regions. The net result of the last process is an enhancement or growth of the crystallite fibrils and the first mentioned reaction causes much of the amorphous portion of the cellulose microstruc-

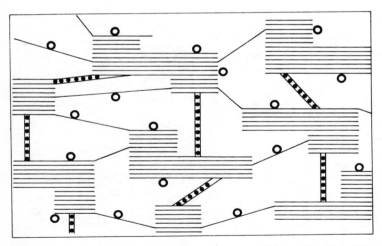

Fig. 3-3. Diagrammatic microstructure of microgranular ion-exchange celluloses. The horizontal solid lines represent the highly orientated regions. The wider dotted lines show the crosslinking introduced into the structure. The dotted circles are ion-exchange sites. (Reproduced with permission from Reeve Angel Scientifica Division, supplier of Whatman advanced ion-exchange celluloses.)

ture to be removed (14, 17). Before ionized groups are attached to these modified structures, the products are crosslinked with bifunctional reagents to restrict the tendency of the cellulose matrix to swell; also the crosslinks prohibit loss of accessibility to the matrix during contraction, since neighboring glucosidic residues are prevented from interacting through H-bond formation (14, 15). The impact of these modifications upon the native structure of cellulose can be seen by comparing Fig. 3-3 with Fig. 3-2.

Relative to an equivalent fibrous exchanger, the ionized groups of a microgranular cellulose exchanger are more accessible to large polyelectrolytes. Furthermore, the modification of the original cellulose structure results in the production of shorter particles that are dense and rod-shaped; this is an advantage in that columns with compact bed material have greater resolving power (14, 17). Although higher nominal capacities (up to 5 meq per g of dry exchanger) are possible in microgranular exchangers, this property is usually only advantageous in the study of small organic solutes, since, as in the case of the fibrous exchanger, there is a tendency towards irreversible retention for large polyelectrolytes if the charge density of an exchanger is too large.

3. Choice of the Exchanger Type

As a result of the original investigations on the preparation and properties of cellulosic exchangers by Sober and Peterson (18), and of ensuing work by Porath (19) and Semenza (20), many different cellulose derivatives are now readily available. Some commercially obtainable ion-exchange celluloses are listed in Table 3-4. These materials have been particularly effective in the purification of serum proteins, nucleic acids, enzymes, and the lower molecular-weight polyelectrolytes that are derived from these macromolecules. The net charge on the material to be fractionated should determine whether a cationic or anionic cellulose exchanger should be employed to purify a given compound. As pointed out by Peterson (17), the net charge on a biologically active polyelectrolyte is not always easily determined. However, simple trial evaluations, with the large solute dissolved in solutions of low ionic strength, and the use of either a cation or anion exchanger should establish the net charge of the polyvalent species. Once this has been determined, the next property to be considered is the acidic or basic strength of the exchanger that will be used for the more extensive treatment of the solute in question. In general, the higher the molecular weight of the polyelectrolyte, the weaker should be the acidic or basic strength of the chosen exchanger. However, due to the complexity of the forces of attraction between a polyelectrolyte and a cellulosic exchanger, general predictions as to the choice of exchanger type cannot always be given. Some general examples of the applicability of cellulose exchangers to the purification of macromolecules are briefly examined

TABLE 3-4

Commerically Available Cellulosic Ion Exchangers[a]

Anion exchangers[b]		Ionizable group	meq/g[c]
AE-cellulose	Aminoethyl-	$-O-CH_2-CH_2-NH_2$	0.3–1.0
DEAE-cellulose[d]	Diethylaminoethyl-	$-O-CH_2-CH_2-N(C_2H_5)_2$	0.1–1.1
TEAE-cellulose	Triethylaminoethyl-	$-O-CH_2-CH_2-\overset{X}{\underset{}{N}}(C_2H_5)_3$	0.5–1.0
GE-cellulose	Guanidoethyl-	$-O-CH_2-CH_2-NH-\overset{NH}{\overset{\|}{C}}-NH_2$	0.2–0.5
PAB-cellulose	p-Aminobenzyl-	$-O-CH_2-\langle\!\!\langle\ \rangle\!\!\rangle-NH_2$	0.2–0.5
ECTEOLA-cellulose	Triethanolamine coupled to cellulose through glyceryl and polyglyceryl chains. Mixed groups		0.1–0.5
BD-cellulose	Benzoylated DEAE-cellulose		0.8
BND-cellulose	Benzoylated-naphthoylated DEAE-cellulose		0.8
PEI-cellulose	Polyethyleneimine adsorbed to cellulose or weakly phosphorylated cellulose		0.1

Cation exchangers[b]		Ionizable group	meq/g[c]
CM-cellulose[d]	Carboxymethyl-	$-O-CH_2-COOH$	0.5–1.0
P-cellulose	Phosphate	$-O-\overset{O}{\overset{\|}{\underset{\underset{OH}{\|}}{P}}}-OH$	0.7–7.4
SE-cellulose	Sulfoethyl-	$-O-CH_2-CH_2-\overset{O}{\overset{\|}{\underset{\underset{O}{\|}}{S}}}-OH$	0.2–0.3

[a]Reproduced from E.A. Peterson, *Cellulosic Ion Exchangers*, American Elsevier Publishing Company, Inc., New York, 1970, by permission of the author and the American Elsevier Publishing Company.

[b]Manufacturers and suppliers: Whatman, manufactured and sold by W. & R Balston (Modified Cellulose) Ltd., Springfield Mill, Maidstone, Kent, England, and available in North America through Reeve Angel & Co., Inc., 9 Bridewell Place, Clifton, New Jersey 070104, U.S.A.; Selectacel, made by the Brown Company, Berlin, New Hampshire, U.S.A., sold under this brand name by the Schleicher and Schuell Company, Keene, New Hampshire, U.S.A.; Serva, manufactured and sold by Serva-Entvicklungslabor, Heidelberg, Germany, and distributed in the United States by the Gallard-Schlesinger Chemical Mfg. Corp., 584 Mineola Avenue, Carle Place, L.I., N.Y. 11514; Cellex, manufactured and sold by Bio-Rad Laboratories, 32nd and Griffin Avenue, Richmond, California, U.S.A., and also available under this brand name from some general biochemical supply companies (information from reference cited in footnote a).

[c]Acid-base capacity as indicated by manufacturer.

[d]Available from Whatman in microgranular form.

here, but more specific information is found in Peterson's book (17). Also given there is practical information pertaining to the packing and operation of cellulosic columns. In the descriptions that follow, see Table 3-4 for details as to exchanger types.

DEAE-cellulose, the most widely used cellulosic exchanger, is a moderately strong base. It exhibits its greatest capacity below pH 10, and has been used extensively for the separation and purification of enzymes, hormones, serum proteins, phospholipids, low molecular-weight nucleic acids (e.g., transfer ribonucleic acids), and oligonucleotides as well as mononucleotides (15, 17).

TEAE- and GE-cellulose are classified as strongly basic exchangers. The latter exchanger, in particular, retains its positive charge at very high pH values (17). Both have been used in chromatography of certain acidic proteins and folic acid derivatives (9).

ECTEOLA-cellulose is a weakly basic exchanger employed for separating viruses, high molecular-weight nucleic acids, nucleoproteins, and large acidic polysaccharides (9, 17).

CM-cellulose, a commonly used weakly acidic cation exchanger, has a high capacity for neutral and basic proteins; it has been used for the fractionation of hemoglobins, albumins, hormones, and enzymes (9).

P-cellulose is a bifunctional cation exchanger. The phosphate groups of P-cellulose are fully ionized above pH 8 but only a single charge exists on each phosphate moiety at pH of 4 or below (17); thus the exchanger contains both strongly and weakly acidic groups. This exchanger, being dibasic, strongly attracts basic proteins (9), and it has also been used frequently to fractionate peptide mixtures (21, 22).

D. ION-EXCHANGE DEXTRANS AND POLYACRYLAMIDE GELS

1. The Gel Matrix

The matrix material of dextran exchangers consists of crosslinked polyglucose chains and the network material of the polyacrylamide exchangers is comprised of a copolymer of acrylamide and methylenebisacrylamide (23). These substances can be compared to the resinous exchangers of Section B in that their three-dimensional networks imbibe water and expand, forming insoluble gel-like structures. When ionogenic groups are introduced into these structures, to form ion-exchange materials, they represent a concentrated solution of electrolyte within the gel structure itself; the added functional groups are all nearly in the interior of these substances. Each gel particle can be considered something like a sponge on a molecular scale.

The dextran and polyacrylamide gel exchangers differ from the resinous exchangers in that the matrix materials of the former are hydrophilic rather than hydrophobic; also the matrices of the dextran and polyacrylamide exchangers have a molecular structure sufficiently open and porous to allow macromolecules to enter and leave.

2. Preparation

Dextran exchangers are prepared by first crosslinking linear polyglucose chains (dextran) with glycerin-ether linkages to give an insoluble hydrophilic carbohydrate network (23) such as that illustrated in Fig. 3-4. Next, functional ionic groups, either acidic or basic and which are similar to some of those contained in cellulose ion exchangers, are attached to the hydroxyl groups of the glucose units of the polysaccharide chains forming, respectively, dextran cation or anion exchangers. Two different dextran anion and cation exchangers are available commercially under the trade name Sephadex; these exchangers are described in Table 3-5.

TABLE 3-5

Commercially Available Dextran Exchangers

Types[a]		Description	Functional groups	Counter ion[b]
DEAE-	A-25	Weakly basic anion	Diethylaminoethyl	Chloride
Sephadex	A-50	exchanger		
QAE-	A-25	Strongly basic anion	Diethyl-(2-hydroxy-	Chloride
Sephadex	A-50	exchanger	propyl) aminoethyl	
CM-	C-25	Weakly acidic cation	Carboxymethyl	Sodium
Sephadex	C-50	exchanger		
SP-	C-25	Strongly acidic cation	Sulphopropyl	Sodium
Sephadex	C-50	exchanger		

[a]The letters A and C are added as suffixes to denote either anion or cation exchanger. They are used in conjuction with the numbers -25 and -50 to designate degree of porosity (see text).

[b]Exchanger form as sold by Pharmacia.

Source: Reproduced from *Sephadex Ion Exchangers*, March, 1971, by permission of Pharmacia Fine Chemicals, Inc., Piscataway, New Jersey.

Polyacrylamide exchangers are prepared in a manner similar to that used for making the dextran derivatives; however, the matrix material is of an entirely different nature. For the acrylamide gel exchangers, water-soluble acrylamide $(CH_2=CH—CONH_2)$ is polymerized in the presence of methylenebisacrylamide

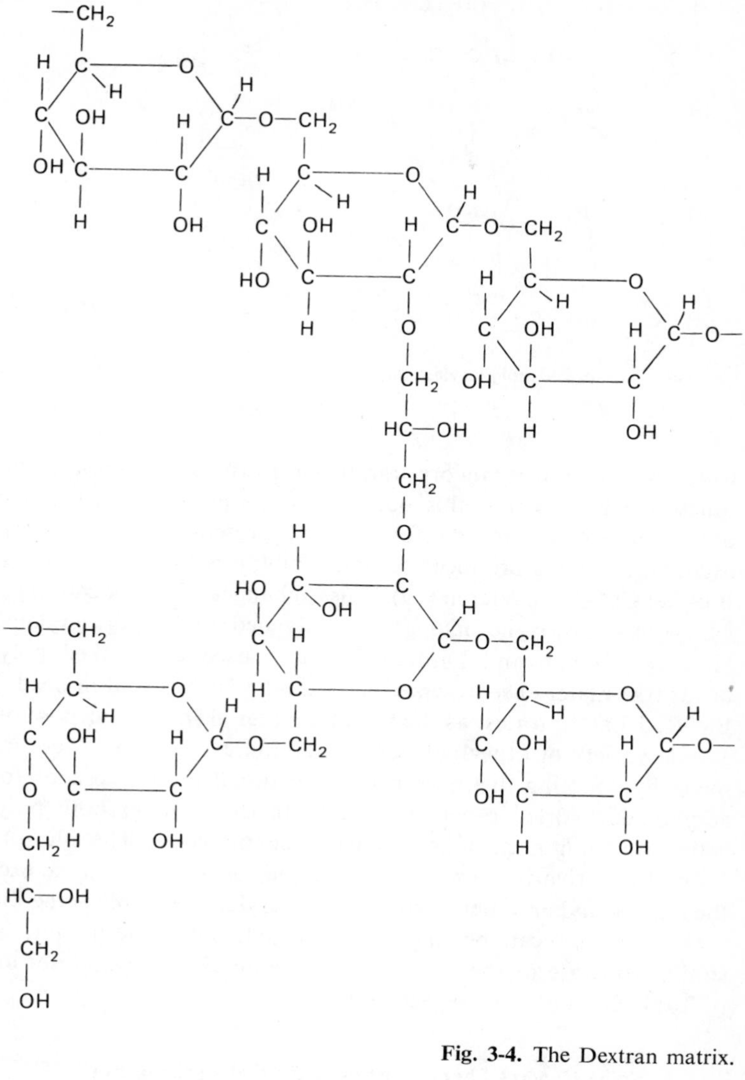

Fig. 3-4. The Dextran matrix.

$$H_2C\!\!=\!\!CH\!-\!\!C\!-\!\!NH\!-\!\!CH_2\!-\!\!NH\!-\!\!C\!-\!\!CH\!\!=\!\!CH_2$$
$$\qquad\quad \| \qquad\qquad\qquad\quad \|$$
$$\qquad\quad O \qquad\qquad\qquad\quad O$$

The presence of the bifunctional reagent causes the formation of crosslinkages between different linear chains of polymerized acrylamide; the end result is a water-insoluble three-dimensional network (23) whose structure is illus-

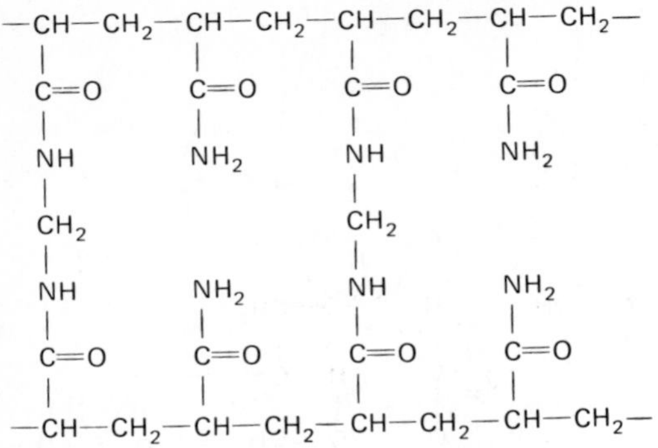

$$-CH-CH_2-CH-CH_2-CH-CH_2-CH-CH_2-$$

Fig. 3-5. Crosslinked polyacrylamide.

trated in Fig. 3-5. By incorporating (on to the amide group) acidic or basic functional groups into this polyacrylamide network, corresponding cation or anion exchangers are formed. At present only one polyacrylamide exchanger is being produced commercially; it is sold under the trade name of Bio-Gel CM-2 (see reference 9). This exchanger is a weak cation gel derivative which has carboxylic acid groups attached to its polyacrylamide matrix. However, Inman and Dintzis (24) have used crosslinked polyacrylamide beads (the material represented in Fig. 3-5 which is sold in bead form by the Bio-Rad Laboratories) as the starting material for the purpose of producing a wide variety of other polyacrylamide ion exchangers. These materials are made by coupling the appropriate functional groups to a preformed polyacrylamide hydrazide derivative (24). In this manner both polyacrylamide cation and anion exchangers have been prepared. The phosphoethyl and sulfoethyl derivatives are typical examples of the former type exchanger and the aminoethyl and diethylaminoethyl acrylamides typifies the kind of anion exchanger that can be made. These exchangers should find applications similar to those of the corresponding ion-exchange celluloses and dextrans of Table 3-4 and 3-5, respectively.

3. Some General Characteristics of the Gel Exchangers (9, 25)

Both the Sephadex and Bio-Gel exchangers can be purchased or prepared in the form of dry spherical beads with a standard particle size in the range of 100–200 mesh. When hydrated, the spherical particles pack easily into columns and give uniform flow rates; since the interstitial volumes are smaller for spherical particles, band spreading and sample dilution are minimized, which aids the ability of the gel exchangers to resolve complex mixtures.

Dextran and acrylamide exchangers are insoluble in water and water solutions. The former exchanger type is not affected by alkaline solutions, but the latter exchanger should not be used in alkaline solutions of pH above 10; both are stable in acidic solutions down to a pH of about 2 (treatment with more acidic solutions can be tolerated for short periods). Exposure to strong oxidizing agents should be avoided for both exchanger types.

In addition to their ion-exchange qualities, the Sephadex and Bio-Gel exchangers exhibit a marked screening effect. In fact the nonionic forms (before functional groups are attached) of these gels are used to fractionate and separate molecules according to their molecular size (this is a technique known as *molecular sieving* or *gel filtration*). There are many different available porosities among the nonionic gels; however, the choice of different porosities among the gel exchangers is limited. Bio-Gel CM-2 is available in only one porosity; it excludes molecules with a molecular weight greater than 10,000. Sephadex exchangers are available in two porosity types designated by the numbers 25 and 50 (see Table 3-5). The highly crosslinked gels of the 25-type have a lower porosity and are used for solutes of molecular weight less than 30,000. The 50-type is more porous and will have the higher available capacity for solutes of molecular weight above 30,000.

For the fractionation of large molecules, the cellulosic exchangers are often preferred to these gel exchangers, since the charge density is greater in the dextran and acrylamide derivatives than it is in the ion-exchange celluloses. As previously discussed, this could lead to instances where macromolecules become very tightly bound. Also the sieving properties of the gel exchangers could prevent some large molecules from penetrating the interior of these gels. Another unfavorable comparison of these gels with the cellulose exchangers is that large changes in bed volumes occur with changes in ionic strength or pH. This could lead to low recovery of material sorbed to an exchanger bed; also the expansion of a gel bed could develop pressures that could break or explode column equipment.

4. Selecting the Type of Gel Exchanger

The charge and molecular size of the solute to be treated will dictate which gel exchanger should be employed. In addition to the degree of crosslinkage, the pore size is influenced by the nature of the counter-ions, the ionic strength, and the pH of the surrounding solution. For practical reasons (e.g., the shrinkage and swelling of gel columns) it is usually advisable to select a gel of the lowest porosity permissible for a given experiment. Since the net charge of a substance should determine whether a cation or anion exchanger should be used, size seems to be the important criterion as to the variety of gel exchangers that should be selected. The pore size of the gel exchanger does not influence the binding mechanism but influences the capacity, since charged groups in the interior of the gels may not be available to large polyelectrolytes. However, compounds of very high molecular weight can be chromatographed on the higher-crosslinked gels but only a fraction of the available capacity will be utilized since only the functional group on the surface of the beads will be available to them (25). In general, the larger the molecule the larger should be the porosity of the exchange material.

Both the dextrans and the acrylamide gels are used to separate the larger natural polymers of biological origin such as acidic, neutral, and basic proteins and the nucleic acids; they are also used to purify and fractionate enzymes, hormones, and are used for the chromatography of lower molecular-weight substances such as peptides, amino acids, and nucleotides. Specific examples of the applicability of the gel exchangers for the fractionation of sensitive organic substances can be found in the references cited for this chapter; other such examples are given in Chapter 7.

E. INORGANIC ION EXCHANGERS

1. The Rebirth of Inorganic Ion-Exchange Materials

Almost in every comparison the resinous ion exchangers have far more superior characteristics than did the earlier natural or synthetic mineral ion exchangers. However, in two specific areas, namely, in the presence of intense radiation fields and at high temperatures, the resinous exchangers possess poor stabilities (3, 6). Thus, with the advent of nuclear reactors and the subsequent development of radiochemical engineering, attention was focused once more on inorganic exchange materials.

Reexamination of the adsorptive properties of inorganic exchange materials was initiated by two different groups roughly at the same time, but independently, at atomic energy installations in Great Britain and in the United States (see references 7 and 8). From the efforts of these two groups, newer inorganic ion-exchange materials with excellent properties were found which are far superior to the zeolites and other older exchange minerals. These investigations at the nuclear plants revealed that certain hydrous oxides, acid salts, heteropolyacids, and insoluble sulfides possessed chemical properties that could compare favorably to the resinous exchangers in regard to their chemical and mechanical stability, insolubility, ion-exchange capacity, ion-exchange rate, and applicability. The ion-exchange crystals are not difficult to prepare and details as to their synthesis can be found in the pertinent references of this chapter; also many of the inorganic exchangers can be purchased from the Bio-Rad Laboratories (9). Some general physical and chemical properties of each of these classes of inorganic exchangers are described.

2. Hydrous Oxide Exchangers*

Studies on a number of insoluble hydrous oxides by Kraus and co-workers (7) and by Amphlett and his co-workers (8) have shown that the hydrous oxides of the tetravalent metals such as zirconium, thorium, titanium, tin, and tungsten may behave either as cation or anion exchangers. The precipitation of a hydrous oxide or hydroxide (e.g., the precipitation of hydrous zirconium oxide with ammonium hydroxide from a solution of zirconyl chloride) represents the final stage in the formation of a crosslinked polymeric network (6, 26, 27). Without the anion of the initial solution or the cation of the precipitating base, these materials are reminiscent of weakly acidic cation exchangers in the hydrogen form or weakly basic exchangers in the free base form; such substances would become charged on treatment with acids or bases (26). Amphlett (27) suggests that either of the following two mechanisms is involved in placing a positive or negative charge on the matrix of a hydrous oxide.

$$\text{Acid solution} \qquad\qquad \text{Alkaline solution}$$

$$\rangle\overset{+}{M} + OH^- \rightleftharpoons \rangle M-OH \rightleftharpoons \rangle M-\overline{O} + H^+$$

$$\rangle M-\overset{+}{O}H_2 \xrightarrow{H+} \rangle M-OH \xrightarrow{OH-} \rangle M-\overline{O} + H_2O$$

Depending upon the basicity of the central metal atom, there is a change from cationic to anionic behavior over a range of pH. At the isoelectric point both cation and anion exchange properties exist and the hydrous oxides are capable of salt pick-up (6, 26, 27).

The hydrous oxides as typically represented by hydrous zirconium oxide show unusually high affinities for polyvalent ions such as sulfate, chromate, phosphate, borate, and carbonate in acid solution where the hydrous oxides behave as anion exchangers (6, 9, 26). The zirconium exchanger can be used to separate the halides and its unique selectivity for fluoride ions is attributed to the known complex-forming reaction between Zr(IV) and fluoride; the hydrous oxide exchanger also is capable of separating certain negatively charged metal compounds such as the chloro complexes of gold and silver (26). The transition to a cation exchanger occurs above the isoelectric point of the hydrous oxides (6, 26, 27). At these higher pH values the negatively charged exchanger attracts exchangeable cations. The cation-exchange behavior of the hydrous oxides has not been investigated as thoroughly as has their anion behavior. However, several unique properties have been noted.

---

*Hydrous aluminum oxide and other previously well characterized materials such as silica are excluded from this discussion.

The cationic hydrous oxides (e.g., hydrous zirconium oxide) exhibit a strong affinity for copper from ammonical solutions and under controlled pH and electrolyte concentration the rare earths may be separated from the alkali metals and the alkaline earths (26). Hydrous titanium oxide shows an unusual order of selectivity in that cesium ions may be eluted before sodium ions (26). A comprehensive review on inorganic ion-exchange chromatography on oxides and hydrous oxides has been compiled by Fuller (28). Structures, selectivities, kinetics, and capacities of these materials are among the topics covered in this review as well as the mechanisms of ion exchange that are associated with these exchangers.

3. Acid Salt Exchangers

The tetravalent metals of the last section also form highly insoluble acid salts when reacted with polyvalent ions such as phosphate, molybdate, tungstate, or arsenate; these salts show marked cation-exchange behavior (6, 26, 27). A typical example from among this class of substances is the microcrystalline gel formed when zirconyl chloride is precipitated with phosphoric acid. The chemical composition of such an exchanger is non-stoichiometric and varies according to the method of preparation; its chemical structure is not known (3, 10). However, evidence has been gathered to indicate that the ion-exchange crystals comprise a matrix with identical repeating units of Zr—O—Zr—O—Zr— to which phosphate groups are attached to the zirconium atom in a containing configuration such as (3, 10)

$$
\begin{array}{ccc}
OPO_3^- & & OPO_3^- \\
| & & | \\
\ldots\ —Zr—O—Zr—O—\ \ldots \\
| & & | \\
OPO_3^- & & OPO_3^-
\end{array}
$$

Similar cation-exchange materials can be prepared from other oxygen-containing acids such as arsenic, molybdic, or tungstic acids and either titanium, tin, thorium, or zirconium can act as the central atom (26, 27).

Acid salts such as typified by zirconium tungstate, zirconium molybdate, and zirconium phosphate possess unique and very useful cation-exchange properties. In their H-form these acid salts show unusually strong affinities for cesium in acidic solutions. In fact, the attraction of cesium for these exchangers under acidic conditions is so great that it can be uniquely isolated from all other elements in the Periodic Table (26). These exchangers also are used to separate cleanly and rapidly the individual member of the alkali and the alkaline earth metal ions; such separations are described in Chapter 7.

4. Heteropoly Acid Salts

Insoluble ammonium phosphomolybdate or phosophotungstate are examples of inorganic cation exchangers which show unusual selectivities towards certain metal ions (6, 9, 27). The former exchanger has been studied in more detail than the other members of this class of materials. It exhibits a strong affinity for K, Rb, and Cs ions but shows very little attraction for sodium and lithium and the Group II cations; all these ions are easily separated on a bed of ammonium phosphomolybdate crystals.

The ion-exchange behavior of these heteropolyacid salts is closely related to the geometrical packing of the large heteropolyanions in the crystal (6, 27). Cations occupy large cavities in the crystal lattice and both these cations and water molecules diffuse freely within these spaces. Exchange of cations between the crystal and a surrounding solution is possible if the entering cations are of the right size to form a stable crystal network with the heteropolyanion. Such restrictions limit the range of cations that can undergo exchange reactions with these materials and also readily explain their unique affinities towards only a few select ions. Multivalent cations are only weakly attracted to the ammonium salts of heteropolyacids in acid solutions but show strong affinities for these exchangers when the pH is adjusted to the neutral range (29).

5. Metal Sulfides

A number of sulfides [e.g., Ag(I), Fe(II), Cu(II), Zn(II), Pb(II), Cd(II), and As(III)] have been found to be excellent adsorbents for many of the transition metal ions (30). In aqueous media, the reactions between the precipitated sulfides and the transition ions is rapid enough so that the process can be carried out in chromatographic columns (30). The reactions appear to be of the sorption-displacement type in which the metal cation of the initial sulfide is displaced or "metathesized" by the cations of the solution whose sulfide is more insoluble.

Cadmium disulfide is an excellent "adsorbent" for heavy metals such as Cu(II), Hg(II), and Ag(I) which form more insoluble sulfides (31). The uptake of these ions by cadmium sulfide can be described by the metathetical reaction:

$$CdS + \left(\frac{2}{Z}\right)M^{+z} \longrightarrow M_{2/z}S + Cd^{++} \qquad (3\text{-}1)$$

where M^{+z} is the displacing ion. The sorption-displacement reaction can proceed to completion, rapidly, under the usual conditions of column operation (41). Other metal sulfides [e.g., Ag(I), Fe(II), Cu(II), Zn(II), and Pb(II)] also enter into these displacement reactions with heavy metals whose sulfides are more insoluble (30). Thus a very select and unique kind of ion exchanger is available for analytical work or for the removal of trace components from a solution.

6. Inorganic Phosphate Gel

Hydroxylapatite (a calcium phosphate gel) is an example of an inorganic material that is used extensively in the biochemical field. This calcium phosphate salt [$Ca_5(PO_4)_3OH$] is used in the fractionation and purification of proteins, enzymes, nucleic acids, and viruses (9, 32). The hydroxylapatites are characterized by their very small crystal dimensions; the resulting large surfaces exhibit ion-exchange properties (32). Because it is of such fine particle size, calcium phosphate gels must be packed into columns with an appropriate filler aid such as diatomaceous earth. Elution of material sorbed to hydroxylapatite is usually carried out with phosphate buffers; neutral salts such as sodium chloride have little effect in removing macromolecules from a bed of phosphate gel (32).

F. MISCELLANEOUS ION-EXCHANGE MATERIALS

1. Some Lesser Used Resinous Exchangers

In addition to the most commonly used resinous exchangers of Table 3-3, there are also several lesser used varieties. Many of these were developed from attempts to produce resins with very select affinities (3, 6). An example of this type of exchanger is the chelating resin Dowex A-1 which has iminodiacetic acid groups attached to a styrene-divinylbenzene matrix; it can be represented by the structure

$$\phi\text{-CH}_2\text{N} \underset{\diagdown \text{CH}_2\text{COO}^-\text{H}^+}{\overset{\diagup \text{CH}_2\text{COO}^-\text{H}^+}{}}$$

Inspection of its formula shows that it could exist in many different zwitterionic forms depending upon the pH. This exchanger forms very stable metal complexes with the transition elements and has much lower affinities for metals which do not undergo complex formation with the iminodiacetic group (3, 9). Due to these properties the chelate exchanger is used for the analysis of trace metals in natural waters, reagents, biochemical and physiological fluids, and to collect traces of metals from process streams. Examples of such applicabilities are given in more detail in Chapter 6. Chelating resins can also undergo ligand-exchange reactions, a process described in Chapter 9.

There are also commercially available resinous exchangers whose acidic or basic strength is intermediate between those listed in Table 3-3. Bio-Rex 63* is a cation exchanger which is an intermediate acid resin; it has phosphonic acid bifunction groups covalently bonded to a styrene-divinylbenzene matrix. An example of an intermediate base resin is Bio-Rex 5,* an anion exchanger, which contains tertiary and quaternary amines on a polyalkyleneamine network.

2. Pellicular Ion Exchangers

These materials consist of a thin superficially porous crust of ion-exchange materials bonded to the nonporous surface of tiny (37–50 μ) glass beads. These exchangers are used in high speed, high pressure chromatographic systems. Since diffusion occurs only in a thin layer on the surface, separations can be carried out 100–1000 times faster if pellicular exchangers are used in place of conventional ion-exchange materials. Several types of either cationic or anionic superficially porous exchangers are available commercially. Their characteristics have been described in a review by Kirkland (33) on modern column packings. Pellicular ion exchangers are particularly useful for the analysis of nucleotides, nucleosides, and of purines and pyrimidines; examples of such separations are given in Chapter 7.

G. ION-EXCHANGE LITERATURE

Publications containing descriptions of ion-exchange methods are so numerous that it is difficult to keep abreast of the newer developments in one's own field let alone have a coherent view of all branches of ion-exchange technology. In the biochemical field, as pointed out by Morris and Morris (4), about half of the work presented in current journals will have some form of chromatographic separation; most of these involve the principles of ion exchange. Fortunately, there are general review type articles and many up-to-date books which appear that contain a wealth of pertinent references that should enable the user of ion-exchange materials to keep informed about any new phase of ion-exchange development.

1. Books

The following textbooks supersede and reinforce the many excellent works that appeared earlier and which are cited together with the following in the chapter references of this book; after each book citation an annotation is given.

*Trade names given to these exchangers by the Bio-Rad Laboratories.

O. Samuelson, *Ion Exchange Separations in Analytical Chemistry*, John Wiley & Sons, Inc., New York, 1963. Specializes in the uses of ion exchangers, particularly the resinous type, in qualitative and quantitative analysis.

W. Rieman III and H. F. Walton, *Ion Exchangers in Analytical Chemistry*, Pergamon Press, New York, 1970. An excellent text for the analytical chemist that provides a broad survey of the role that ion exchange can play in chemical analysis.

F. Helfferich, *Ion Exchange*, McGraw-Hill Book Company, Inc., New York, 1962. Essentially a treatise on ion-exchange theories; it contains practical information but most of the subject matter is considered from an advanced mathematical viewpoint.

E. Heftmann (ed.), *Chromatography*, Reinhold Publishing Corporation, New York, 1967, 2nd ed. A book not exclusively devoted to ion exchange but includes a description of all chromatographic materials; it contains a large collection of references.

E. A. Peterson, *Cellulosic Ion Exchangers*, American Elsevier Publishing Company, Inc., New York, 1970. A book describing the ion-exchange celluloses and their uses in the biochemical field; it contains very practical information.

J. J. Kirkland, (ed.), *Modern Practice of Liquid Chromatography*, Wiley-Interscience, Inc., New York, 1971. This book deals with the theory, equipment, and column packings associated with high pressure, high performance chromatographic systems. Ion exchange is considered among other chromatographic processes.

P. R. Brown, *High Pressure Liquid Chromatography—Biochemical and Biomedical Applications*, Academic Press, New York, 1973. This book describes the fundamentals of theory and instrumentation of high pressure liquid chromatography and demonstrates the application of high performance chromatography to all branches of biology, chemistry, medicine, and related fields. This book should serve as a basic reference source for those now using high pressure liquid chromatographic techniques and also should serve as an introductory text for those unfamiliar with the modern practices in liquid chromatography.

J. A. Dean, *Chemical Separation Methods*, Van Nostrand Reinhold Co., New York, 1969. A teaching text, for chemistry majors, that emphasizes the newer separation procedures. Extensive coverage is given to ion-exchange methods. The book contains clearly presented tables and figures.

2. Handbook

Handbook of Chromatography, Vols. I and II (edited by Gunter Zweig and Joseph Sherma), published by The Chemical Rubber Co., Cleveland,

Ohio, 1972. This handbook should serve as a working manual and reference book for investigators in all fields of chromatography. Volume I contains over 549 tables that give chromatographic data (e.g. column size, packing, solvent, flow rate, temperature, detection, and literature sources) for a wide variety of compounds. Volume II is designed to give investigators, even those inexperienced in chromatography, a practical insight into the theory and practices of ion exchange. The handbook also describes other chromatographic processes.

3. Reviews

Several journals (not specifically devoted to ion-exchange work) publish a special issue yearly or on alternate years that review exclusively only the ion-exchange work for a certain period. They are:

Annual Review of Physical Chemistry, "Ion Exchange," issued yearly starting 1951.

Analytical Chemistry, Annual Reviews, "Ion Exchange," issued on alternate years beginning in 1948.

Industry and Engineering Chemistry, Annual Review, "Unit Operations," Ion-Exchange Section, issued yearly beginning in 1948.

Other review articles appear in book form with each volume containing several articles, each of which specializes or emphasizes different types of chromatography. These books are:

Advances in Chromatography

Chromatographic Reviews

4. Journals

Some journals are exclusively devoted to the publication of articles that deal with chromatographic procedures, materials, or equipment. These are:

Journal of Chromatography

Journal of Chromatographic Science

Other journals such as *Analytical Biochemistry* are not exclusively devoted to chromatographic work but each issue usually contains several articles that emphasize chromatographic techniques.

When one evaluates the preceding literature sources, it is understandable that the individual worker must depend more and more upon the review-type article for his information rather than attempt to read the enormous numbers of original publications.

REFERENCES

1. Kunin, R., *Elements of Ion Exchange*, Reinhold Publishing Corporation, New York, 1960, Chap. 2.

2. Wolf, F. J., *Separation Methods in Organic Chemistry and Biochemistry*, Academic Press, Inc., New York, 1969, Chap. 5.

3. Helfferich, F., in *Advances in Chromatography*, Marcel Dekker, Inc., New York, 1965 (ed. by J. C. Giddings and R. A. Keller), Vol. 1, Chap. 1.

4. Morris, C. J. O. R., and Morris, P., *Separation Methods in Biochemistry*, Interscience Publishers, Inc., New York, 1963, Chaps. 8, 10.

5. Sober, H. A., and Peterson, E. A., in *Ion Exchangers in Organic and Biochemistry*, Interscience Publishers, Inc., New York, 1957 (ed. by C. Calmon and T. R. E. Kressman), Chap. 16.

6. Salmon, J. E., *Progress in Nuclear Energy*, Series *IX Analytical Chemistry*, Pergamon Press, Oxford, 1961, Vol. 2, p. 338.

7. Kraus, K. A., Carlson, T. A., and Johnson, J. S., *Nature*, **177**, 1128 (1956); Kraus, K. A., and Phillips, H. O., *J. Am. Chem. Soc.*, **78**, 694 (1956); *J. Am. Chem. Soc.*, **78**, 249 (1956).

8. Amphlett, C. B., McDonald, L. A., and Redman, M. J., *Chem. & Inds.* (London), 1314 (1956); *J. Inorg. Nucl. Chem.*, **6**, 220 (1958); **6**, 236 (1958).

9. Bio-Rad Laboratories, *Materials for: Ion-Exchange-Gel Filtration-Adsorption*, Richmond, Cal., June 1971.

10. Helfferich, F., *Ion Exchange*, McGraw-Hill Book Company, Inc., New York, 1962, Chap. 2.

11. Samuelson, O., *Ion Exchange Separations in Analytical Chemistry*, John Wiley & Sons, Inc., New York, 1963, Chaps. 2, 8.

12. The Dow Chemical Company, *Dowex::Ion Exchange*, Midland Mich., 1964 Chap. 1.

13. Mikes, O., *Chromatographic Methods*, D. Van Nostrand Company, Ltd., London, 1961 (chief ed. O. Mikes and translation ed. R. A. Chalmers), Chap. 5.

14. Knight, C. S., in *Advances in Chromatography*, Marcel Dekker, Inc., New York, 1965 (ed. by J. C. Giddings and R. A. Keller), Vol. 4, Chap. 3.

15. Reeve Angel Scientifica Division, *Whatman Advanced Ion-Exchange Celluloses* (*Data Manual and Catalog*-2000), Clifton, New Jersey.

16. Keller, R. A., and Giddings, J. C., in *Chromatography*, Reinhold Publishing Corporation, New York, 1967 (ed. by E. Heftmann), 2nd ed., p. 128.

17. Peterson, E. A., *Cellulosic Ion Exchangers*, American Elsevier Publishing Company, Inc., New York, 1970 (ed. by T. S. Work and M. Work), Chaps. 2–5.

18. Sober, H. A., and Peterson, E. A., *J. Am. Chem. Soc.*, **76**, 1711 (1954); Peterson, E. A., and Sober, H. A., *J. Am. Chem. Soc.*, **78**, 751 (1956).

19. Porath, J., *Arkiv Kermi*, **11**, 97 (1957).

20. Semenza, G., *Helv. Chim. Acta*, **43**, 1057 (1960).

21. Anfinsen, C. B., Sela, M., and Cooke, J. P., *J. Biol. Chem.*, **23**, 1825 (1962).

22. Epstein, C. J., Anfinsen, C. B., and Sela, M., *J. Biol. Chem.*, **237**, 3459 (1962).

23. Determann, H., *Gel Chromatography*, Springer-Verlag, New York, 1968, Chap. 2.

24. Inman, J. K., and Dintzis, H. M., *Biochemistry*, **8**, 4074 (1969).

25. Pharmacia Fine Chemicals, Inc., *Sephadex-Ion Exchangers*, Piscataway, N. J., March 1971.

26. Kraus, K. A., Phillips, H. O., Carlson, T. A., and Johnson, J. S., *Proc. 2nd Intern. Conf. on Peaceful Uses of Atomic Energy.*, **28**, 3 (1958).

27. Amphlett, C. B., *Inorganic Ion Exchangers*, Elsevier Publishing Co., New York, 1964, Chaps. 4 and 5.

28. Fuller, M. J. *Chromatog. Rev.*, **14**, 45 (1971).

29. Smit, J. Van R., Robb, W., and Jacobs, J. J., *J. Inorg. Nucl. Chem.*, **12**, 95 (1959); **12**, 104 (1959); see also Smit, J. Van R., and Robb, W., *J. Inorg. Nucl. Chem.*, **26**, 509 (1964).

30. Phillips, H. O., and Kraus, K. A., *J. Chromatog.*, **17**, 549 (1965).

31. Phillips, H. O., and Kraus, K. A., *J. Am. Chem. Soc.*, **85**, 486 (1963).

32. Levin, O., in *Methods in Enzymology*, Academic Press, Inc., New York, 1962, Vol. V, p. 27. (ed. by S. P. Colowick and N. O. Kaplan),

33. Kirkland, J. J., *Anal. Chem.*, **43**, 37A (1971); see also Kirkland, J. J., in *Modern Practice of Liquid Chromatography*, Wiley-Interscience, Inc., New York, 1971 (ed. by J. J. Kirkland), Chap. 5.

Laboratory Columns and Accessories: Operational Techniques

FOUR

A. THE CHROMATOGRAPHIC ASSEMBLY

A basic block diagram for a multipart chromatographic system is shown in Fig. 4-1. There can be extreme variations in the individual parts that make up such an assembly. A very simple arrangement can be put together, in a few minutes, from the common glassware and apparatus found in the ordinary laboratory; in this case, the various devices of the assembly usually are

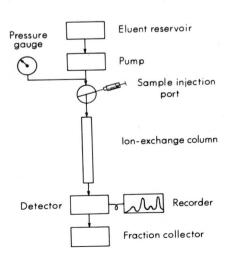

Fig. 4-1. Block diagram for a multipart chromatographic system.

operated manually. More elaborate systems consist of expensive, complicated modules that can be made to operate completely or semi-automatically.

For difficult separation work, a more complete assembly is almost always used, while for simpler fractionations some of the devices of Fig. 4-1 can be omitted. Each part of the chromatographic system and its operation or function is discussed in the sections that follow.

B. CHROMATOGRAPHIC COLUMNS

1. Column Shape; Length to Diameter Ratio

The packed column is the principal component of the chromatographic assembly; a cylindrical tube is the most suitable design for such a column. There are no particular advantages (such as sharpness and evenness of solute bands) to be gained by packing an exchanger into a square, funnel-shaped, conical, or any other form different from the cylinder (1).

If it were not for imperfections in the packed column, disturbances upon the application of a sample, and the spreading of a solute band during its migration, the dimensions of the cylindrical column would be only of minor importance. But since these operational flaws do occur in practice, the shape (i.e., long and narrow versus short and wide) of a column is of major importance; experience has taught that the bed height of an exchanger should be greater than its diameter.

The ratio of length to diameter of an exchanger bed depends upon the type of operation for which the packed column is intended and to some extent on the type of exchanger itself. A common inefficient practice is to use a column longer than necessary. Even for the resolution of several similar components in a mixture (e.g., sugars, mononucleotides, or the alkali metals), the length of the exchanger bed need only be 10 to 20 times its diameter. This is only a rule of thumb for routine work since for certain separations, particularly in the biochemical field, it is not uncommon for the length to be 50 to 100 times the column diameter. Columns with these latter dimensions usually contain the ion-exchange low capacity celluloses, or dextrans which are used principally for the separation of high molecular-weight solutes. The length-to-diameter ratio is decreased when higher-capacity materials, such as the resinous or inorganic types, make up the exchanger bed. Sharp resolutions can be achieved on these exchangers in columns in which the bed height seldom exceeds 20 cm or the internal diameter 1 cm. Although for analytical work long and narrow columns are used, the same chromatographic system can be expanded for preparative work by simply keeping the height constant and increasing the diameter of the column to accommodate any convenient scale of operation. In these cases, as the cross-sectional area of a column increases, the flow rate used to process a given volume of solution may also be increased proportionally.

For less stringent chromatographic work such as desalting, removal or recovery of trace components from a large volume of solution, conversion of an exchanger from one form to another, separation of an ion having a strong affinity for an exchanger from an ion having only a weak affinity, or a determination of total salt concentration, it is not uncommon to use columns in which the length-to-diameter ratio approaches unity.

In some cases, the selection of a column with the proper dimensions to carry out a particular operation will necessarily involve trial and error experimentations; ordinarily, reference to the literature will provide the required information or at least point the way to a good starting point for empirical testing, or provide sound theoretical considerations.

2. Column Types

The chromatographic column, in its simplest form, is a tube with a support inserted at the bottom end that should retain the ion-exchange material above it but allow liquid to pass through the column. The support can be a fritted disc made from glass, metal (preferably stainless steel), a circular piece of porous plastic (such as Teflon or polyethylene), a mat of glass wool or asbestos covered with a layer of small diameter glass beads, or sieve-like materials such as rigid netted Nylon or Saran. The tube itself can be made of glass, metal, or plastic; such materials should be inert to the solvent chosen

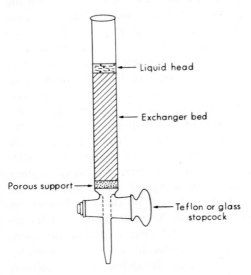

Fig. 4-2. A burette converted to an ion-exchange column.

for a particular chromatographic operation. To allow for temperature control, the columns can be jacketed so that heated or cooled liquids may be pumped around the outer surface of the chromatographic tube. Due to their various cross-sectional sizes and lengths, burettes serve as a ready supply for chromatographic columns of different diameters and capacities. They can be converted to chromatographic columns by inserting, just above the stopcock, one of the various bed supports just mentioned. Such a column is shown in Fig. 4-2.

More elaborate columns of various lengths and diameters can be fabricated from glass tubing or glass piping. Such columns can be constructed so that in addition to the affixed porous supports, stopcocks and connector devices such as ball or tapered joints are part of the apparatus. These refinements are convenient during the packing of the ion-exchange materials and allow the columns to be more easily connected to other parts of the chromatographic assembly. Some typical designs for these types of columns are shown in Fig. 4-3.

Professionally designed columns of various dimensions are available from chromatographic supply houses. There are several advantages to be gained through the purchase of commercially obtainable equipment. Many of the marketed columns are constructed so that they: (1) are of precision bore throughout their entire length; (2) are able to withstand specified pressures; (3) are easily jacketed to allow for temperature control; (4) have fittings at both the entrance and exit ends of the column that can be secured or loosened rapidly and conveniently; (5) have column supports that are readily accessible for replacement or cleaning and whose removal aids in the repacking of a column; (6) have essentially no "dead space" between the top of the exchanger bed and the entrance tube that feeds solution to the bed material. Some typical examples of commercially available chromatographic columns and assemblies are shown in Fig. 4-4(a) through Fig. 4-4(c).

3. Packing the Column

Most exchangers imbibe water (or other polar solvents) and swell considerably when brought into contact with aqueous solutions. Therefore, to prevent a column from cracking or exploding and for personal safety, an exchanger first should be allowed to expand, outside the column, in a large excess of water contained in a wide diameter vessel such as a beaker.

In some cases, before introducing the ion-exchange material into a column, it may be necessary to pretreat the exchange material further. This usually involves removing unwanted "fine" particles and also trace impurities that have been retained by the exchangers as a result of their manufacture. Fine particles may be removed by decantation after an exchanger-slurry mixture has been allowed to settle; repetition of this process will remove the majority of the fine particles. Removal of impurities can also be accomplished by a batch process by stirring the exchanger with chosen solvents. After each equilibration the exchanger is retained on a filter (e.g., a sintered glass Buchner type funnel); the exchanger is then placed in fresh solvent and the

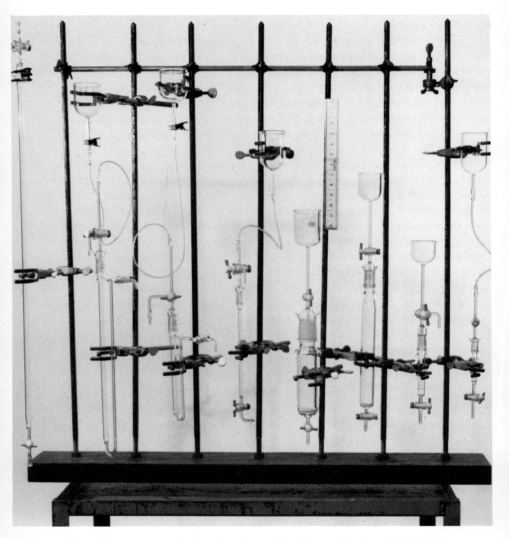

Fig. 4-3. Homemade columns of various designs. Proceeding from left to right, the approximate inside diameters (in cm) are: 1.3, 1.0, 1.0, 0.6, 4.6, 3.1, 1.5 and 1.0. (Photograph courtesy of Oak Ridge National Laboratory.)

Fig. 4-4(a). Component parts for attaching the bottom end piece to a Bio-Rex column: Solv-seal joint (a); plastic threaded column collar (b); porous support disc (c); detachable column tip (d); top connection is the same, only the porous disc is omitted. (Photograph courtesy of the Bio-Rad Laboratories.)

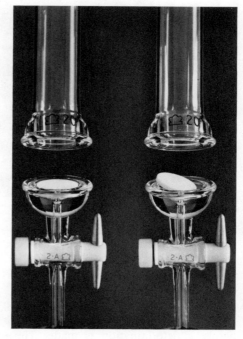

Fig. 4-4(b). The bottom connection for a Kontes chromatographic column. The porous support discs are available in glass or plastic (e.g., polythylene). The end piece is attached to the column with a compression clamp after inserting an "O" ring. The top connection is the same except the support disc is omitted. (Photograph courtesy of the Kontes Glass Company.)

**Analytical
Column Assemblies**

13 x 0.9 cm . . .

23 x 0.9 cm . . .

29 x 0.9 cm . . .

69 x 0.9 cm . . .

169 x 0.9 cm . . .

**Preparative
Column Assemblies**

60 x 1.9 cm . . .

60 x 3.8 cm . . .

160 x 1.9 cm . . .

160 x 3.8 cm . . .

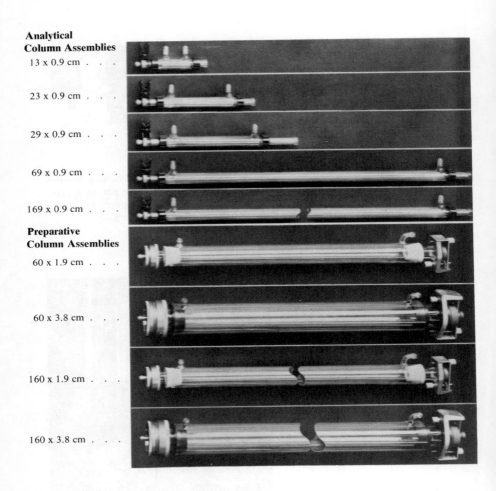

Fig. 4-4(c). Beckman complete column assemblies. The columns themselves are borosilicate glass cylinders whose inner diameters vary only ± 0.0005 of an inch throughout the entire length of the glass body. (Photograph courtesy of Beckman Instruments, Inc.)

process is repeated. The last solvent employed is often that solution to be used to slurry the exchanger into a column and to equilibrate the column before a determination is made. The solvents utilized depend upon the chemical properties of the exchange material under treatment; such properties are discussed in Chapter 3. If analytical grade exchangers are available, the pretreatment steps can be kept to a minimum number, the main step being that of preswelling the exchange material.

Several different techniques can be used to pack a preswollen exchange material into a column. Before following any of the methods given here, the empty column should be backwashed (a backward flow from the exit end of the column) in order to remove air from the bed support and to have liquid below and above it.

(a) Multistage batch-packing

1. Drain the column to about $\frac{1}{8}$ to $\frac{1}{4}$ of its length (the distance above the bed support to the top of the column).

2. Add the exchanger as a slurry (about 1 volume of settled exchanger to 1 volume of water) to fill the column again.

3. Allow the exchanger to settle in the column.

4. Drain the liquid down or aspirate from above so that a few centimeters of liquid are still above the settled bed of the exchanger.

5. Repeat steps (1) through (4) until the column is filled to the desired height.

(b) Single-stage batch-packing (see Fig. 4-5)

1. Fill the column to the top with water.

2. Attach a funnel via an adapter to the top of the column (the diameter of the funnel stem and adapter should be approximately equal to that of the column).

3. Position a stirring motor above the funnel.

4. With the stirrer on add a slurry (1 volume of settled exchanger to 1 volume of solution) of exchanger (the volume of the exchanger should be greater than that needed to pack the column) to the funnel.

5. Let the liquid drain from the column; if possible, the flow rate at the exit end should be equivalent to or faster than the linear rate at which the exchanger bed is being settled.

(c) Single-stage pump-packing

1. In reference to Fig. 4-6 a slurry reservoir is positioned over the chromatographic column. The top chamber should have a larger volume (for convenience it may have a larger diameter) than the column to be packed.

2. The bottom column is filled with solution.

3. An excess of exchanger is added as a slurry to the slurry reservoir.

4. Liquid from a pump is forced into the slurry reservoir.

5. Exchanger particles are forced into the chromatographic column at a velocity much greater than their settling rate.

6. Pumping is continued until the bottom column is fully packed.

The choice as to the method of packing depends upon the physical properties of the exchanger under investigation. As particles become smaller and/or the exchange materials less dense, the job of packing them into a

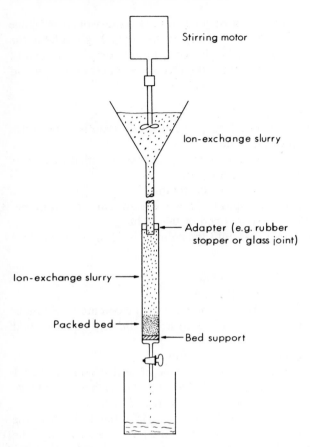

Fig. 4-5. Gravity flow arrangement for single-stage batch-packing of an ion-exchange column.

column becomes more difficult. Most routine work with the resinous exchangers is carried out with a particle size of 200–400 mesh or larger; in these instances packing method (*a*) is used. For work requiring smaller particle size, where higher resistance to column flow is encountered, method (*c*) is utilized. When extremely high pressures are to be encountered, as in the dynamic packing of very fine particles (e.g., resin beads about 20 μ in diameter), high pressure equipment (e.g., see reference 2) is used in place of the apparatus illustrated in Fig. 4-6. Method (*b*) is generally used to pack the ion-exchange celluloses and the ion-exchange dextrans. Methods (*a*), (*b*), or (*c*) can be employed to form an exchanger bed of inorganic ion-exchange crystals. In some cases asbestos is mixed (1 : 1, W : W) with the inorganic exchangers to facilitate packing. The choice of method depends upon the size of the inorganic aggregates.

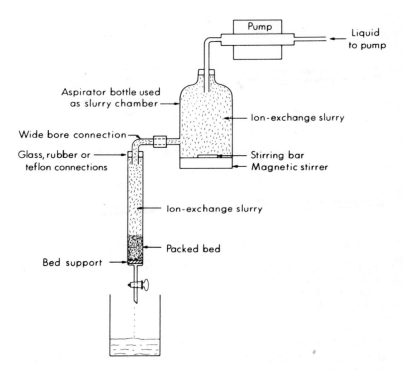

Fig. 4-6. Single-stage, pump-packing arrangement for filling an ion-exchange column.

C. LINE CONNECTIONS FROM ONE ACCESSORY TO ANOTHER

If the complete chromatographic assembly of Fig. 4-1, or only parts of it, are used, there must be a continuous flow of liquid from one piece of equipment to the next. To avoid changes in concentration due to mixing, all piping leading from one accessory to another should be of narrow bore; inside diameters should be on the order of 2–3 mm or less.

Glass, metal, rubber, or plastic tubing is employed to make the line connections from one device to another. Stainless steel, heavy-walled glass tubing, or pressure-rated plastic tubing are found mainly in the systems that are operated at high pressures (> 10 psi). At lower operating pressures, such as in gravity feed systems, flexible tubing made from rubber, polyvinyl chloride, polyethylene, polyurethane, or Teflon are common piping materials; capillary glass tubing is often used in combination with these materials. Even in the assemblies operated at high pressure, some line connections may be of flexible plastic, rubber, or ordinary glass tubing, since

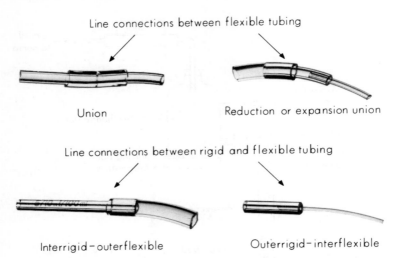

Fig. 4-7. Commonly used "sleeve" couplings for joining two lengths of tubing or for connecting tubing to chromatographic equipment. Tight seals may be made among the same materials or between different materials by choosing the proper diameters and wall thicknesses.

high pressure exists only between the exit end of the pump and the column bed support. Before and after these points there is essentially free flow of liquid.

Since flexible tubing is available in a variety of inside diameters and wall thicknesses, it is a simple matter to join two lengths of tubing together, even though they may be of different diameter or of different material. A few such common type couplings are shown in Fig. 4-7. Connections of the tubing (e.g., the end pieces) to the individual devices of the chromatographic assembly may also be made by these same types of unions. In addition to the couplings illustrated in Fig. 4-7, various threaded connectors (e.g., unions, reducers, tees, valves), of metal or plastic, are available commercially (see Fig. 4-8). Many of these fittings are designed especially for chromatographic work.

D. APPARATUS FOR DELIVERING THE ELUENT TO THE COLUMN

Eluting solutions are delivered to an ion-exchange column by one of two different procedures. In the first of these, the *single-solvent* method, there is no change in the composition of the eluent as it is supplied to the column. In the other procedure, the eluent is delivered to the column in

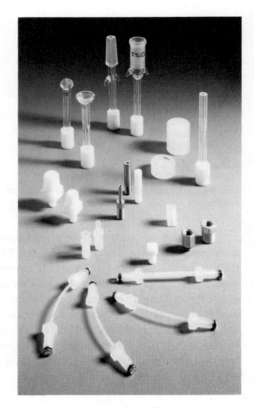

Fig. 4-8. Some typical commercial type fittings designed for connecting chromatographic equipment. (Photograph of Cheminert fittings courtesy of Chromatronix Incorporated.)

such a fashion that it *does* change in composition during the elution process. This latter method is known as *gradient elution.*

For feeding a solution of constant composition to the column, the elution reservoir is usually a vessel having only a single chamber, although for the convenience of increasing the capacity, two or more vessels containing the same solution may be interconnected. A few devices that are used for this method of elution are illustrated in Fig. 4-9.

Eluents of changing composition are delivered to the column by any one of the several different types of gradient devices. Bock and Ling (3) and also Snyder (4) have reviewed the principles of gradient elution and have evaluated some of the various designs for elution reservoirs that are applicable for this technique. The presentation of this subject is limited here, and only a few of the simpler designs for gradient elution vessels are illustrated.

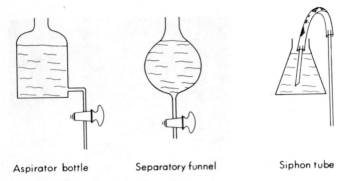

Aspirator bottle   Separatory funnel   Siphon tube

Fig. 4-9. Apparatus for delivering solutions of constant composition to the column.

Designed variations in the eluent composition can be brought about by a *discontinuous* or a *continuous* process. The first mentioned operation is better known as *stepwise elution* and is carried out by supplying to the column successive changes of different solutions so as to increase the eluent strength at each step. This procedure is performed manually by emptying each previous solution from a reservoir before introducing the next. By having the solutions of increased eluting power in different containers, each with its own stopcock or valve, but connected to a common manifold [see Fig. 4-10(a)], stepwise elution can be programmed on a time basis by the activation of the

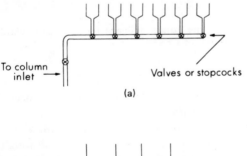

To column
  inlet   Valves or stopcocks

(a)

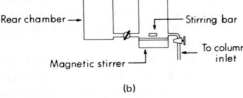

Rear chamber   Stirring bar

Magnetic stirrer   To column
                                 inlet

(b)

Fig. 4-10. Apparatus for delivering solutions of changing composition to the column: (a) delivers a discontinuous or stepwise gradient; (b) delivers a continuous gradient.

proper valve at the proper interval. If a large number of changes are made, the increase in eluent strength can be gradual, but due to the inherent limitations of this technique it is employed, mainly, just to give a small number of sharp increases in the eluting power of the eluent.

A smooth continuous rise in eluent power is achieved simply by the manner in which the eluting reagent is delivered to the column. Although more complicated devices are often used, the working of the two vertical-walled gradient vessels of Fig. 4-10(b), first described by Parr (5), demonstrates the principle of how a solution undergoes continuous changes in composition as it is being supplied to a column. The two reservoir containers of the figure just mentioned are connected by a stopcock, or by flexible tubing; a stirrer is positioned in the front chamber which contains a dilute solution and has a small diameter exit tube that leads directly to the column. A more concentrated solution (or one of greater eluent strength, such as one consisting of a different solvent or a different pH) is placed in the rear chamber. As flow begins under continuous stirring, small portions of the more concentrated solution are mixed immediately with a large volume of the less concentrated solution. As a result the solution leaving the exit tube of the gradient device undergoes a smooth rise in eluent power. If the densities of the two solutions are similar, the concentration of liquid emerging from the Parr gradient device is given by

$$C_v = C_2 - (C_2 - C_1)\left(1 - \frac{v}{V_t}\right)^{A_2/A_1} \tag{4-1}$$

where C_v is the concentration of the delivered volume v, A_1 and A_2 are the cross-sectional areas of the reservoirs that initially contain solutions of concentration C_1 and C_2, respectively, and V_t is the total volume initially present in both containers (3).

As seen in Fig. 4-11, if both chambers of the Parr apparatus are equal in cross-sectional area, a linear rise of concentration is observed, but if one vessel differs in area from the other, a nonlinear rise in concentration is obtained. This may be either steep or gradual as noted by the relationships given in Fig. 4-11.

Some gradient elution reservoirs are designed with as many as nine separate compartments. These multichambered containers have more flexibility in producing gradients than do the simpler devices. A change from a steep to a gradual gradient or vice versa, or even back and forth from a linear to an exponential change of eluent power can be achieved during the delivery of solution from these multicompartment devices.

The choice as to which type of elution apparatus to include in the chromatographic assembly depends upon the complexity of the separation work. If the separation can be achieved efficiently by elution with a solution of constant composition, then there are a few definite advantages to be gained.

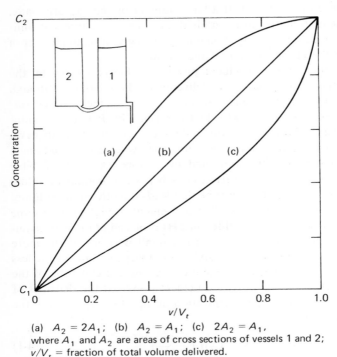

(a) $A_2 = 2A_1$; (b) $A_2 = A_1$; (c) $2A_2 = A_1$,
where A_1 and A_2 are areas of cross sections of vessels 1 and 2;
v/V_t = fraction of total volume delivered.

Fig. 4-11. Graphical representation of concentration gradients as delivered by the Parr apparatus. (Reproduced from R. M. Bock and N. S. Ling, *Anal. Chem.*, **26**, 1543 (1954), by permission of the authors and the American Chemical Society.)

First of all, there is simplicity in the apparatus, and secondly, the operation of equilibrating the column anew with starting solution following each analysis is eliminated. The chief drawback to the single-solvent systems is that in the latter part of the chromatogram, elution bands are broadened, poorly shaped, and require excessive elution times. These unwanted qualities can be eliminated by increasing the eluent strength during the course of an elution. Thus the objective of gradient elution is to reduce the retention time of solutes strongly sorbed to an ion exchanger, yet at the same time to increase the overall resolution of the solute bands as they move down the column. These aims have been illustrated schematically by Snyder (4) for a hypothetical separation. As shown in Fig. 4-12, improvement in elution profiles occurs when gradient elution is used to elute the components rather than solutions of constant composition; as noted in the figure, the gradients may be stepwise or continuous.

The eluent strength chosen for a continuous gradient is determined by

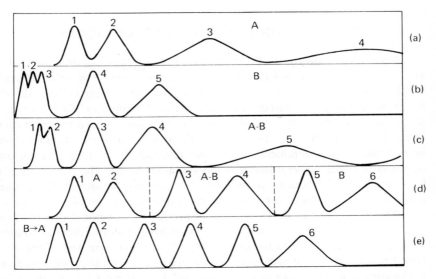

Fig. 4-12. Comparisons of elution patterns in fixed elution and gradient elution chromatography. A, B, A–B, and B ⟶ A represent eluent strengths (A is weak, B is strong): (a) hypothetical separation of a six-component mixture in fixed eluent elution chromatography, weak eluent; (b) same, strong eluent; (c) same, eluent of intermediate strength; (d) stepwise gradient elution of same mixture; (e) gradient elution of same mixture. (Reproduced from L. R. Snyder, *Chromatog. Rev.*, **7**, 1 (1965), M. Lederer (ed.), Elsevier Publishing Company, Amsterdam, 1965, by permission of the author and the Elsevior Publishing Company.)

trial and error. During this exploration period, to find the optimum gradient, stepwise elution can be exploited to its fullest advantage. By this technique, unlimited changes of eluent strength can be easily made in response—or lack of it—to solute concentrations in the effluent; thus if material is being eluted slowly or not at all, a solution of greater eluent power can be introduced to the elution reservoir (6). In this manner, desired elution patterns can be achieved from small volume inputs with the result that only a relatively few fractions need to be analyzed. From such experiments, the optimum conditions for constructing a continuous gradient are easily deduced.

Stepwise gradients are very seldom recommended for difficult separation work. Since this gradient method lends itself to sudden changes of solvent, which result in the production of sharp elution bands, spurious peaks (band splitting) may occur, or more than one solute may appear in a single peak. Furthermore, frequent changes of eluent can become laborious. For these reasons stepwise gradients are utilized mainly in preliminary work, or when the components of a mixture have such different affinities for the exchange material that their elution bands are widely separated.

E. DETECTION DEVICES AND TECHNIQUES.

General considerations   Precise location of separated solute bands is the key to accurate analysis by chromatographic methods. For column work involving the collection of only a few fractions, one simply collects the effluent manually in containers such as beakers, graduates, flasks, or test tubes, and determines by a suitable method the solute concentration in each of the individual fractions. However, as the separation work becomes more refined and many components are present in a sample mixture, it is not uncommon to collect several hundred individual fractions. The collection of a large number of fractions is necessary to properly construct the *elution curve*, namely the plot of the solute concentration versus the effluent volume.

Almost any test or device capable of producing a response to a physical or chemical property of substances may be utilized to detect them in column effluents. Depending upon the complexity of the separation and of the analytical results desired, detection of solutes may be a simple process, such as performing a manual spot test (i.e., the $AgNO_3$ test for Cl^-) on successive portions of the effluent, or detection may involve automatic devices that continuously monitor the column effluent and that produce a tracing of the elution patterns on a moving-chart recorder. In this latter method of detection a necessary part of such monitoring systems is a flow-through cell suitably designed to measure a chemical or physical property of the flowing solution. As pointed out by Morris and Morris (7), the design of flow-through cells is by no means straightforward. The cell has to be constructed so as to minimize turbulence and unswept areas, so that an accurate record of the solution compositon entering the cell can be made. As a result of thorough experimentations [e.g., see the references cited by Morris and Morris (7)] highly sensitive, low dead volume, microliter capacity, flow-through cells that proficiently measure the optical, electrical, or radioactive properties of solutes as they appear in column effluents are now commercially available. When such a cell assembly detects the presence of solute, the response to the chosen physical or chemical property is converted electronically to a signal that, upon amplification, subsequently transmits the magnitude of the response to a moving-chart recorder; in this manner elution curves are mechanically plotted. The amount of solute represented by a peak of the automatically drawn elution curve is related to the area of that individual peak. Quantitation of elution curves is discussed in Chapter 5.

Good detecting systems exhibit a high sensitivity (the ratio of detector response to the amount of sample) above a low background (or noise) response (8). It is generally assumed that a substance is "detectable" if the sensitivity response signal is at least twice the noise (9). In addition, the detector system should have a wide linear-response range, that is, the quantity being measured should be linearly proportional to changes of solute concentration over a wide range of values. These concepts are illustrated diagrammatically in Fig. 4-13.

Methods of detection may be classified into those procedures that measure the optical (e.g., ultraviolet or visible spectral absorption, fluorescence, refractive index), electrical (e.g., electrical conductance, electrochemical potential, dielectric constant), biological (e.g., enzyme specificity, growth stimulation or inhibition), or nuclear (e.g., radioactivity) properties of solutes (7, 8, 9, 10). Measurements of electrical conductance, index of refraction, or dielectric constant are examples of nonspecific or bulk analytical procedures, while measurements of light absorption, radioactivity, oxidation-

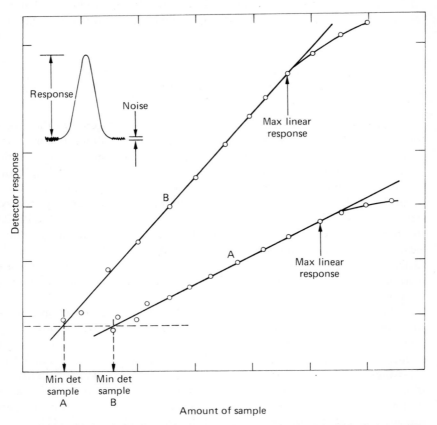

Fig. 4-13. Hypothetical detector response as a function of sample size; samples A and B are of different sensitivities. (Reproduced from M. N. Munk, in *Basic Liquid Chromatography*, Varian Aerograph, Walnut Creek, Calif., 1971, by permission of the author and Varian Aerograph.)

reduction potential, or enzyme specificity are examples of selective or specific detection methods. Nonspecific detection procedures are usually less sensitive and are more prone to interferences from changes of temperature and eluent strength (e.g., from gradient elution systems) than are specific detection methods.

Some analytical procedures that are, in principle, applicable for chromatographic work are useless on a practical basis because they do not have the required sensitivity to detect solutes in low concentration (e.g., in biochemical work, solute concentrations may be in the picomolar range). However, there are many analytical procedures that are compatible with ion-exchange work. Detection techniques as used in column liquid chromatography have been reviewed by Byrne (9), Snyder (11), Huber (12), Conlon (13), Munk (8, 14), and Veening (15). For specifics dealing with the detection of solute bands in eluate fractions the reader is directed to the literature sources just mentioned.

F. CONTROL OF LIQUID FLOW THROUGH
THE COLUMN ASSEMBLY

One of several different techniques may be used to control the rate at which the eluting solution is forced through the column and its component parts. Flow by gravity is the most simple and convenient way of admitting eluent to a column. In this case rate of flow is controlled by the height that eluent reservoirs, such as those shown in Fig. 4-9, are placed above a column. When lack of height becomes the limiting factor for gravity feed systems, the rate of liquid flow through the column assembly may be controlled by regulating the pressure of an inert gas above the eluent in the elution reservoir. Neither of the two nonmechanical methods, just described, for delivering eluent to a column are ideally suited for very precise control of solvent flow; this is especially so in discontinuous or continuous gradient elution.

Mechanical pumping offers an efficient, precise, compact manner of forcing the eluting solution through an exchanger bed. Variable flow pumps are commercially available that are capable of operating with a flow constancy of 2–3% at flow rates between 0–10 ml per min against column back pressures that range from 2–5000 psi (11, 16, 17, 18). Precise control of flow rates is important in high resolution work, since uncontrolled changes in eluent flow affect retention time, the degree of resolution, and can cause undesirable changes in the baseline of flow-sensitive detectors (17).

Low pressure column work (0–50 psi) may be carried out with variable flow peristaltic (tubing pumps), centrifugal, piston, solenoid pulsating, or Saturn type pumps (descriptions are given in supplier's catalogs). In high resolution work, wherein the particle size of the exchanger may be 10–20 μ, columns may be operated at inlet pressures up to 5000 psi. Specially designed reciprocating piston, reciprocating diaphragm, or pneumatic pumps are required for these extremes of pressure (16, 17, 18).

Before pressurizing column equipment, precaution should be taken to consider just what pressure the chromatographic apparatus will withstand. Ordinary glass tubing can withstand a pressure up to about 10 psi. At pressures exceeding this value, glass, metal, or plastic equipment especially rated for high pressure work should be used in that part of the chromatographic assembly that is subjected to the inlet pressure. The effluent emerging from the column (beyond the bed support) is usually only at atmospheric pressure, regardless of the inlet pressure.

G. FRACTION COLLECTORS

Only when a continuous monitoring device gives a quantitative analytical result, and the solutes themselves are of no intrinsic value, is effluent containing separated solute zones discarded without collection. When a flow-through detector is not used or when it only gives a qualitative result, exact solute concentrations in each separated zone have to be determined in individually collected fractions. Manual collection is convenient only when a few fractions need to be collected. In most other situations, automatic fraction collecting devices are an essential, integral part of the chromatographic assembly.

The automatic fraction collector, in simplicity, is a mechanized rack that allows a series of containers (usually test tubes) to position themselves, one at a time at predetermined intervals, under the exit stream of a chromatographic assembly. Basically there are two different ways of delivering effluent to the containers of these mechanical fraction collectors. In one type of collector, the effluent stream is passed through a length of flexible hose which is attached to a device that is guided laterally, at predesignated intervals, across a row of stationary test tubes. After the last tube in a row is filled, the device returns to its original starting position and at the same time an empty row of tubes is pushed forward one position to take the place of the preceding group. In the other type of collector, the exit tube from column apparatus remains in a stationary position over an effluent container and after a fixed time interval a fresh container will replace the preceding one. In this latter case, if the rack is of circular construction, after one revolution the exit tube may be moved automatically to become stationary again over a fresh concentric row of tubes or containers.

Succeeding empty effluent containers may be moved one at a time into a collecting position by: (1) variable electrical timers that may be preset

Fig. 4-14(a). Spiral type volumetric fraction collector. The spiral consists of polyethylene snap tubes that are hinged together into long flexible chain lengths (100–400 tubes). The tubes are disposable or, if desired, can be washed and reused; also glass test tubes may be inserted into the snap tubes. A light beam shining through the tubes detects the height of liquid and thus measures the volume (0.5 ml to 14.5 ml) predetermined for each tube. This fraction collector may be operated in the cold. (Photograph courtesy of Gilson Medical Electronics, Inc.)

Fig. 4-14(b). Assembly line type of fraction collector. By attaching additional extension tracks an unlimited number of racks may be placed into position to collect fractions; each rack holds forty test tubes (18 mm). This fraction collector counts drops or measures volume photoelectrically for any preset period of time. The instrument can operate at temperatures as low as −20°C. (Photograph courtesy of Brinkman Instruments, Inc.)

to trigger the collector at fixed time intervals; (2) constant volume devices which, upon filling, activate the collector; (3) drop-counters set to initiate a signal after a predesignated number of drops have been counted; (4) manual means through an electrical impulse generated by mechanically activating a set of contact points.

Automatic fraction collectors of different designs and sizes are obtainable commercially. The models shown in Fig. 4-14(a) through (c) illustrate three basic modes of operation of collecting devices, namely, the spiral, the linear, and the circular types of fraction collectors. These instruments are ideally suited for collecting fractions having a volume of 25 ml or less. Other models (not illustrated) have racks that accommodate larger containers and are used in low resolution work or when only one or two components may appear in the effluent. Some fraction collectors have self-contained refrigeration units that will keep the collected fractions cold, other models may be operated in a cold room for this purpose.

Fig. 4-14(c). Circular rack type of fraction collector. Each concentric row will accommodate 50 tubes. Delivery tube changes to another row at the end of one revolution. Fractions may be collected on the basis of drop counting or time. The instrument may be operated in cold rooms. (Photograph courtesy of Technicon Instruments Corporation.)

H. OPERATION OF THE CHROMATOGRAPHIC ASSEMBLY

Many ion-exchange procedures require, in the way of apparatus, only a simple liquid reservoir, the packed column, and a few containers to receive the effluent. However, due to the impact of modern instrumentation, the trend in the practice of liquid chromatography has been shifted to more complex automated methodology. In this regard, the following operational adjustments pertain mainly to a multicomponent chromatographic assembly, but where applicable also to a simpler system.

After packing the column (see Section B) the passage of several column volumes of initial eluent through the chromatographic assembly allows:

1. Chemical equilibration of the exchanger bed and thermal equilibration of both the column material and the column apparatus.

2. The chosen flow rate to be established by either adjusting the pumping rate in pressurized assemblies or the height of the reservoir in gravity flow systems.

3. The base line or background level to be established in the detecting apparatus.

4. The fraction collector to be set for the proper time or volume interval between changes of effluent containers.

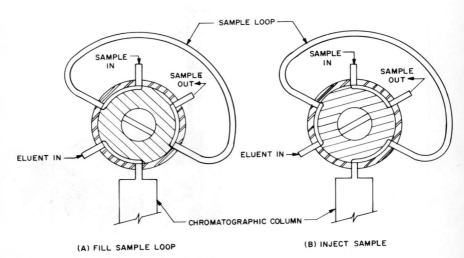

Fig. 4-15. Use of a six-post valve to inject a sample into the eluent stream of the column. (Reproduced from C. D. Scott, in *Modern Practice of Liquid Chromatography*, J. J. Kirkland (ed.), Wiley-Interscience, Inc., New York, 1971, by permission of the author and Wiley-Interscience, a division of John Wiley & Sons, Inc.)

Upon completing the foregoing start-up procedure, the sample is next placed into the effluent stream. Large-sample volumes may be put on stream by interrupting liquid flow momentarily, to replace eluent above the exchanger bed with liquid sample material, after which flow is resumed, or by adding eluent to a reservoir immediately after the sample disappears from it as it is drained into the column. Small-sample volumes may be placed on stream by forcing them (e.g., with a syringe; see Fig. 4-1) into the column via an injection port or valve (17, 18, 19). The latter technique is used routinely to add samples to pressurized columns which have essentially no liquid space above the exchanger bed. Large-volume samples are sorbed to pressurized columns by means of a loop type injector (18, 19). These devices (see Fig. 4-15) allow the flow rate and operating pressure to be sustained while the sample is forced into the column. In all sample loading operations, precaution should be taken to avoid introducing bubbles or air into the eluent stream. Liquid should be always maintained above the exchanger bed.

REFERENCES

1. Cassidy, H. G., in *"Technique of Organic Chemistry, Adsorption and Chromatography,* Interscience Publishers, Inc., New York, 1957 (ed. by A. Weissberger), Vol. 10, Chap. VII.

2. Scott, C. D., *J. Chromatog.*, **42**, 263 (1969).

3. Bock, R. M., and Ling, N. S., *Anal. Chem.*, **26**, 1543 (1954).

4. Snyder, L. R., *Chromatog. Rev.*, **7**, 1 (1965).

5. Parr, C. W., *Biochem. J., Proc.*, **56**, xxvii (1954).

6. Peterson, E. A., *Cellulosic Ion Exchangers*, American Elsevier Publishing Company, Inc., New York, 1970 (ed. by T. S. Work and M. Work), Chap. 2.

7. Morris, C. J. O. R., and Morris, P., *Separations Methods in Biochemistry*, Interscience Publishers, Inc., New York, 1963, Chap. 14.

8. Munk, M. N., in *Basic Liquid Chromatography*, an in-house publication of Varian Aerograph, Walnut Creek, Calif., 1971, Chap. 6.

9. Byrne, Jr., S. H., in *Modern Practice of Liquid Chromatography*, Wiley-Interscience, Inc., New York, 1971 (ed. by J. J. Kirkland), Chap. 3.

10. Polesuk, J., *American Laboratory*, Part 1, p. 27, May 1970; Part 2, p. 37, June 1970.

11. Snyder, L. R., in *Chromatography*, Reinhold Publishing Corporation, New York, 1967, (ed. by E. Heftmann), 2nd ed., pp. 93–96.

12. Huber, J. F. K., *J. Chromatog. Sci.*, **7**, 172 (1969).

13. Conlon, R. D., *Anal. Chem.*, **41**, 107A (1969).

14. Munk, M. N., *J. Chromatog. Sci.*, **8**, 491 (1970).

15. Veening, H., *J. Chem. Education*, **47**, No. 9, A549 (1970); No. 10, A675 (1970); No. 11, A749 (1970).

16. Henry, R. A., in *Modern Practice of Liquid Chromatography*, Wiley-Interscience, Inc., New York, 1971 (ed. by J. J. Kirkland), Chap. 2.

17. MacDonald, F., in *Basic Liquid Chromatography*, an in-house publication of Varian Aerograph, Walnut Creek, Calif., 1971, Chap. 5.

18. Berry, L., and Karger, B. L., *Anal. Chem.*, **45**, No. 9, 819A (1973).

19. Hadden, N., in *Basic Liquid Chromatography*, an in-house publication of Varian Aerograph, Walnut Creek, Calif., 1971, Chap. 2.

Quantitation of Elution Curves

FIVE

A. ANALYSIS OF INDIVIDUAL FRACTIONS

Often in chromatographic analysis, the effluent is collected as a series of discrete fractions, and in these instances the quantitation of elution peaks consists merely of summing up the solute content as found for each of these individual fractions (or for several adjacent fractions lumped together). This technique is practiced when the effluent is not automatically monitored or when a monitoring system does not give a desired quantitative result. Thus, following the actual chromatographic separation, the analysis is completed only when the solute concentrations of the collected fractions are determined subsequently in a further analytical operation.

The results of this type of chromatographic analysis, especially when two or more components are present in a mixture, are best visualized and interpreted from the elution curve, i.e., a graph of the solute concentration found for each fraction versus the accumulative eluate volume. There are two ways of plotting the elution curve. If a large number of fractions of constant volume are collected for each peak, one may identify each fraction by its container number and then plot this number on the abscissa versus the concentration, as determined for that fraction, on the ordinate. As shown in Fig. 5-1(a), a smooth curve may then be drawn through the points, each of which is in close proximity to its neighbor. If the volumes of the collected fractions vary with respect to each other, the elution curve is best plotted as a histogram. In this case, as illustrated in Fig. 5-1(b), the actual volume span of each fraction is represented on the abscissa and the average solute con-

111

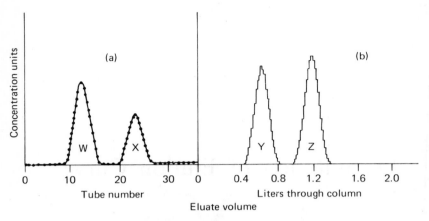

Fig. 5-1. Hypothetical elution graphs: (a), the separation of W and X represented by smooth curves; (b), the separation of Y and Z represented by histogram plots.

centration found for that volume span is represented on the ordinate; a smooth curve is sometimes drawn through the midpoints of each step of the histogram, but this is inaccurate when the number of fractions per peak is less than, say, three.

B. ANALYSIS OF AUTOMATICALLY PLOTTED ELUTION CURVES

Quantitation of automatically recorded elution peaks involves a two-step process. Firstly, the area under the elution curve must be determined and secondly, the area once known must be related to the amount of sample placed on the column. In the discussions that follow it is assumed the amount of sample material appearing in the effluent is identical to that placed on the column.

1. Methods of Obtaining the Area Under Elution Peaks

The area under an elution curve may be determined by several different methods. The most sophisticated methods involve voltage integrating devices attached to a recorder. These devices are capable of electronically integrating an elution curve as it is being drawn by the recorder pen. The more elaborate voltage integrators are interfaced to digital printers which output peak positions as well as peak areas. Such automatic integrators are discussed by Baumann (1) and are not considered further here; however, some less sophisticated and more readily applicable methods of integration are described in the following sections.

(*a*) *Geometrical integration*   This technique is commonly employed and gives a very precise way of obtaining the peak area (1, 2, 3, 4). The method is based on the fact that normal peaks approximate a triangle, hence, the simple formula for obtaining the area of a triangle, namely area $= \frac{1}{2}$ base $\times$ height, can be used to calculate the approximate area under a symmetrical elution curve. There are two ways of constructing a triangle from an elution curve.

The preferred method (see Fig. 5-2) calculates the area by the relation: area $(A) =$ peak height $(H) \times$ width (W') at half peak height. In this case the apex of the triangle is located at C_{max} (the midpoint of the elution curve at height H) and from this point lines are drawn through the set of points $\frac{1}{2} H$ that are located on either flank of the elution curve; the lines intersect the base line and form a triangle whose area coincides with most of the area of the elution curve. Anderson (2) has shown that the area calculated by this procedure (i.e., the $H \times W'$ method) is about 6 % too low compared to what is the true area under a Gaussian elution curve, and uses the absolute correction factor, 1.064, to correct the former area to the latter.

A less preferred method (see Fig. 5-3) uses the triangle drawn tangentially to the elution curve through the inflection points on curve located at $He^{-1/2}$.

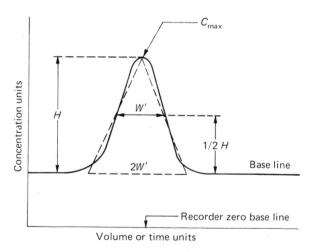

Fig. 5-2. Integration of an elution curve by the peak height $(H) \times$ width (W') at half peak height method. By this procedure an isosceles triangle is drawn from three points common to an elution curve. The apex of this triangle is located at the concentration maximum (i.e., C_{max}) which is the midpoint of the elution curve at peak height, H. From the apex, the two equal sides of the triangle are drawn through the set of points, $\frac{1}{2}H$, that are located on either flank of the elution curve; peak width between these points is the distance W'. The sides intersect the base line to form the isosceles triangle whose base is $2W'$. The area (A) of the elution curve is calculated from the equation: $A = H \cdot W'$. Note that W' is not equivalent to W of Eq. 5-10.

Area is calculated by the relation: $A = \frac{1}{2}$ base (B) × height (H') where B is the distance between the base-line intercepts of the tangents and H' (equivalent to $2He^{-1/2}$) is the height from B to the apex (which, in this case, lies above the midpoint of the elution curve) of the constructed triangle. This procedure is subject to more error since it is often difficult to draw the tangents at their precisely proper position which, in turn, makes it difficult to determine the true H' or B of the desired triangle. When integrated properly, the area measured in this manner gives about 97 % of the true area (1).

If the ordinate values of elution curves, such as those shown in Fig. 5-2 or Fig. 5-3, are plotted logarithmically, the peak height must be calculated in actual scale units as represented on the elution graph. However, if the ordinate scale is linear, peak height and width may be measured in any convenient arbitrary units such as millimeters or inches when calculating peak areas. Neither of the triangular methods for calculating area are accurate if the elution curves are nonsymmetrical.

   (*b*) *Other lesser used methods*   Two other methods sometimes used to integrate elution peaks are planimetry and the actual weighing of an elution

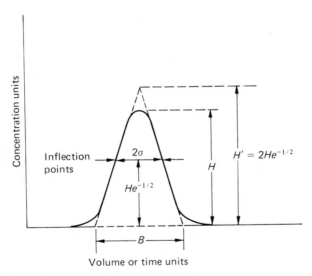

Fig. 5-3. Integration of an elution curve by the $\frac{1}{2}$ base (B) × height (H') method. By this procedure a triangle is drawn tangentially to an elution curve through the inflection points on the curve which, as shown in the figure, are located at $He^{-1/2}$; peak width between these points is the distance 2σ. This triangle has a base equal to 4σ (equivalent to the distance B which lies between the base-line intercepts of the tangents) and an altitude, H', equal to $2He^{-1/2}$, where H is the peak maximum of the elution curve. The area (A) of the elution curve is calculated from the equation: $A = \frac{1}{2}B \cdot H'$.

profile (1). The former is one of the least accurate and least precise ways to integrate a peak. In this case, the area is determined by tracing the perimeter of an elution profile with a planimeter, a mechanical device that gives perimeter distance in digital form on an indicator; nonreproducibility of the manual operation is the main drawback in the use of this technique. The weighing of an elution "cut-out" profile can give very reproducible numbers in terms of the "weight" of a given area of the paper. For better accuracy and consistent results, the elution profiles are preferably photocopied onto paper of uniform thickness and moisture content. Photocopying also preserves the original elution profile.

2. Conversion of Peak Area to Amount of Solute

Once the peak areas of an elution profile have been determined by any chosen method, the next operation is to convert these area values into data that reveal sample composition. As noted in Chapter 2, the area under an elution curve is proportional to the total amount of solute that produces that curve. The exact relation between these two quantities is fundamental to quantitative chromatography.

(*a*) *Direct relation of peak area to sample composition*   If the solutes as represented on a chromatogram have the same response to a given detector, i.e., equimolar amounts of different solutes yield the same area value, then no conversion factors are needed in order to calculate sample composition. Thus if A, B, and C are three solutes having the same response to a detector but are present in a sample in different amounts, as shown in Fig. 5-4, then the percentage of A, for example, is given by

$$\%A = \frac{\text{area of A} \times 100}{\text{area of (A + B + C)}} \qquad (5\text{-}1)$$

However, more often the situation arises where the response of a given detector is different for each different solute. In this case individual conversion or response factors are needed before sample composition can be evaluated (1, 3, 4, 5).

(*b*) *Conversion of peak area into weight or moles of substance through response factors*   For a given detection system without prior calibration, it is difficult to predict how large the area of elution curve will be corresponding to a solute of known weight. Therefore response factors are routinely obtained by chromatographing samples of known composition. Individual peak areas for each known solute are determined from these standard chromatographs by one of the methods previously described. A response factor

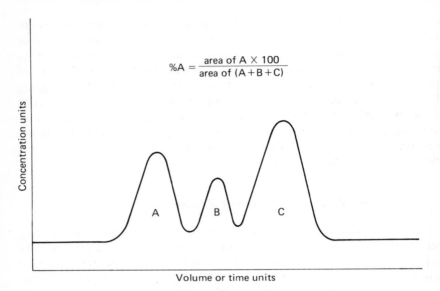

$$\%A = \frac{\text{area of A} \times 100}{\text{area of (A + B + C)}}$$

Fig. 5-4. Conversion of peak areas into per cent composition when each solute gives the same equimolar response to a detector. A, B, and C are represented as three such solutes present in a mixture in different amounts. Sample composition is calculated as shown in the figure and as explained in the text.

F is then calculated for each standard by the expression

$$F_s = \frac{S}{A_s} \tag{5-2}$$

where S represents the amount of a particular standard added to the column and A_s is the area of that solute as found by integrating its peak. Depending upon the units used to represent S, factors are obtained that give the response of a detector in terms of such expressions as milligrams per unit area or moles per unit area, etc. Consequently, in the analysis of unknown samples, peak areas determined from a chromatogram are multiplied by their proper response factors in order to obtain the amount of each solute in the sample added to the column. Thus in the analysis represented in Fig. 5-5, the three solutes have different response factors and the percentage of Y, for example, in the sample is found from

$$\%Y = \frac{A_y \cdot F_y \cdot 100}{A_x \cdot F_x + A_y \cdot F_y + A_z \cdot F_z} \tag{5-3}$$

where $A_x F_x$, $A_y F_y$ and $A_z F_z$ give a "corrected area" so that the different molar responses of the different solutes are considered when evaluating sample composition.

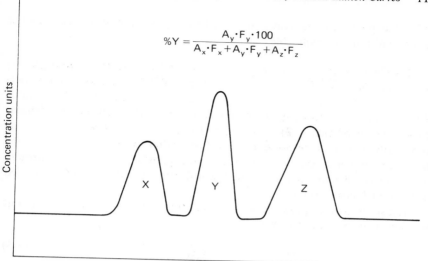

$$\%Y = \frac{A_y \cdot F_y \cdot 100}{A_x \cdot F_x + A_y \cdot F_y + A_z \cdot F_z}$$

Volume or time units

Fig. 5-5. Conversion of peak areas into per cent composition when each solute present in a mixture gives a different molar response to a detector. X, Y, and Z are represented as three such solutes. In this case, as different from the analysis of Fig. 5-4, peak areas are multiplied by a proper response factor to give a "corrected area" that is used in calculating sample composition as is shown in this figure and as explained in the text.

Under very precise control of operating conditions, especially in regard to flow rate and temperature, response factors need only be given in terms of peak height. As pointed out by Uziel *et al.* (6), under identical conditions and at low sample loadings, the peak width of a substance at half peak height will remain constant regardless of the height of a peak, i.e., regardless of the amount of substance added to the column. Therefore, under stringent control of operating conditions, so that the peak of a substance is always at the same elution position, a response factor may also be calculated from

$$F_s = \frac{S}{H_s} \qquad (5\text{-}4)$$

where S is the amount of a standard and H_s is its peak height as determined from a standard chromatogram. Such factors are used in the same manner as the area-response factors of Eq. 5-2 to calculate sample composition. Although the peak height method of calculating sample composition is simpler, the area method is more widely used in such computations because this method is much less sensitive to small variations of the operating conditions. In other words, if a change in peak position causes a change in peak height, a corresponding and opposite change in peak widths will occur and the area under the peaks will remain constant; hence predetermined area-response factors will remain applicable.

(*c*) *Conversion of peak area into quantity of substance through extinction coefficients*   The areas of elution peaks detected spectrophotometrically and subsequently integrated by the $H \times W'$ or the $\frac{1}{2} B \times H'$ methods may be converted into weight or moles of substance through the use of molar absorptivity coefficients (2). This method eliminates the need to determine empirically individual response factors. From what follows it should become clear that the conversion of peak areas into quantity of substance through extinction coefficients can apply to the different operating conditions used in different laboratories.

If a single elution peak is collected in a single container, the total amount of solute may be obtained from

$$\mu\text{moles solute} = \frac{\mu\text{g}}{\text{MW}} = \frac{A \cdot \text{ml}}{\epsilon \cdot l} \tag{5-5}$$

where A is the absorbance (optical density) at a given wavelength in a cell

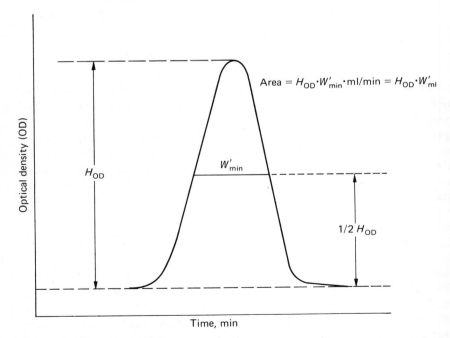

Fig. 5-6. Geometrical integration of an elution peak detected spectrophotometrically. Here the abscissa scale is given in time units, therefore, the width at $\frac{1}{2}H_{\text{OD}}$ in volume units, namely, W'_{ml} is obtained from $W'_{\text{min}} \times$ flow rate (ml/min).

of l-cm path length and ϵ is the extinction coefficient in A units per μmole at the wavelength of measurement (2). If the same peak is drawn automatically on a strip chart recorder via the electrical impulses received from a flow-through detector cell, then, integrating by the $H \times W'$ method (see Fig. 5-6), it follows that

$$\mu\text{moles solute} = \frac{\mu g}{MW} = \frac{H_A \cdot W'_{min} \cdot \text{ml per min}}{\epsilon \cdot l} \tag{5-6}$$

where the net peak height (H_A) is in A units, the width (W'_{min}) at half peak height is in minutes, ml per min is the flow rate of the eluate and ϵ and l are defined as given in Eq. 5-5. If the area is corrected by the Anderson (2) factor and the result expressed for μmoles, then for a cell of 1-cm light path

$$\mu\text{moles} = \frac{H_A \cdot W'_{ml} \cdot 1.064}{\epsilon} \tag{5-7}$$

where W'_{ml} is *always* expressed in *volume* units.

C. CONSTANCY OF OPERATING CONDITIONS

1. Preliminary Considerations

The requirement that liquid flow through the chromatographic system be maintained at a constant and known rate is essential for reproducible and accurate results by any of the quantitation methods (see Section F of Chapter 4). Particularly when elution curves are automatically recorded, elution positions are often given in terms of time instead of volume. Thus, in these cases, the chromatographic identification rests upon how well a given retention time (time to peak) can be reproduced. The convenience of operating at a constant known flow rate also eliminates the need for frequent measurement of the liquid flow rate through a column. In addition, uncontrolled changes of flow rate may affect the degree of resolution found for a given chromatographic condition. Another important factor in quantitative chromatography is the reproducibility of detector response, i.e., the same amount of solute should always give the same net response from a detector.

2. Internal Standards

A check on the constancy of operating conditions such as flow rate, detector response, or any other suspected changes in column conditions and the calculation of a factor that will correct for some of these changes, may be made by either of two methods: by routinely "running" a standard chromatogram every so often, or by frequently including an *internal standard* in the samples that are to be analyzed. The use of the former method is straightforward. A performance check by the latter method involves a comparison of the peak elution time and of the areas (or peak heights) obtained for a standard that is always added in indentical amounts to each sample before analysis. The area of the internal standard as initially determined is preferably obtained from the same chromatograph that yields the data from which the response factors of Eq. 5-2 or Eq. 5-4 are calculated. If the initial area of the standard is A_{is} and the area found in a later chromatogram, if changed, is A'_{is} the factor A_{is}/A'_{is} may be used to normalize or correct the area of an unknown found in that same later chromatogram. Thus the percentage of, for example, the component Y in a sample is found from

$$\%Y = \frac{A_y \cdot F_y}{W_s} \cdot \frac{A_{is}}{A'_{is}} \cdot 100 \tag{5-8}$$

where A_y and F_y are terms previously defined and used in Eq. 5-2 or Eq. 5-3 and W_s is the weight of sample taken for analysis. The internal standard should elute each time in the same position (i.e., have a reproducible retention time) about halfway through the chromatogram, be resolved from other peaks of interest, and must not be present in samples subject to analysis (1, 5).

D. PRECISION AND ACCURACY OF THE INTEGRATION METHODS

A few examples in the literature illustrate the degree of precision and accuracy that is possible for analysis as carried out by chromatographic methods. Integrating peaks with a precision of $100 \pm 3\%$ by the $H \cdot W'$ method, Spackman et al. (3) have used response factors to detect, quantitatively, micromole quantities of amino acids via the ninhydrin color reaction; after the addition of sample their system operated automatically. This technique was refined by Hamilton (4) who detected amino acids at the 0.01 μmole level. Uziel et al. (6) and Singhal and Cohn (7), employing automated systems, have used the $H \cdot W'$ extinction coefficient method to quantitatively (accuracy $\pm 3\%$) detect nanomole quantities of nucleosides found in nucleic acid hydrolyzates. The cut-and-weigh method of integration can yield peak areas to within a precision less than $\pm 2\%$ but is the least convenient and ranks high in the time to perform the integration; planimetry and triangulation may be reproduced to within $\pm 5\%$, while voltage integrating devices are capable of being precise to within $\pm 0.5\%$ and of performing the integration the fastest (1).

E. RESOLUTION

Neither the separation factor, as defined by the ratio of the distribution coefficients (see Eq. 1-23 of Chapter 1) for two components, nor the difference between the peak elution volumes $\bar{v}$ (or the difference between their respective volume distribution coefficient D_v) of two components (see Eq. 2-30 of Chapter 2) give a quantitative measure of resolution. Under a chosen set of conditions, peak separation or differences in distribution coefficients for a pair of components is directly related to the *selectivity* of the column, that is, the inherent potential the exchanger has which allows it to thermodynamically distinguish one component from another. The plate theory predicts that if there is a finite difference in distributions coefficients for two components, however small, the distance between peak maxima of the elution curves will increase, in direct proportion to the length of the column, as the solute zones move down the exchanger. However, the plate theory predicts also that the width of solute zones will increase, in proportion to the square root of the bed length, as they travel down a column. Thus, although solute bands separate faster than they spread, the situation may arise where the peak maxima of two components of similar selectivity are distinctly separated yet the solute zones remain cross-contaminated due to the overlapping of the back portion of one elution curve with the front portion of the other. Therefore, any meaningful or exact measure of separation must take into consideration the width of adjacent elution curves as well as the distance between their peak maxima. Such quantitative evaluations of adjoining elution curves is termed *resolution* (R) which in reference to Fig. 5-7, is defined by the equation (see 8, 9, 10, 11, 12, 13)

$$R = \frac{\bar{v}_b - \bar{v}_a}{2\sigma_a + 2\sigma_b} = \frac{D_v'' - D_v'}{2\sigma_a + 2\sigma_b} = \frac{\Delta d}{2\sigma_a + 2\sigma_b} = \frac{2\,\Delta d}{W_a + W_b} \qquad (5\text{-}9)$$

where $2\sigma_a$ (or $2\sigma_b$) is the peak width between each set of inflection points (located at $C_{max}e^{-1/2}$) for the Gaussian-shaped elution zones. Peak width at the base line of such zones is equal to $4\sigma_a$ (or $4\sigma_b$) and is that distance (equivalent to W_a or W_b of Fig. 5-7) between the base-line intercepts of the tangents to the inflection points.

All the terms of the numerator and the denominator of Eq. 5-9 must be expressed in dimensionally consistent units such as volume or time. When Δd is equal to $2\sigma_a + 2\sigma_b$, adjacent 2σ band widths will just touch (see Fig. 5-7) if projected along the base line toward each other from each peak maxima. In this case $R = 1$ and separation is considered analytically complete (8, 10, 13), but since the elution curves do not conform overall to the configuration of a triangle the Gaussian elution zones will overlap to the extent

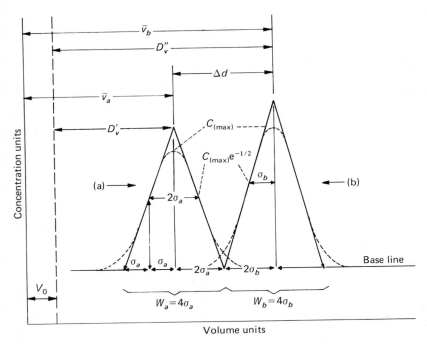

Volume units

Fig. 5-7. Terms that are used to calculate resolution are schematically illustrated. As shown here and explained in the text, the two elution peaks, (a) and (b), are separated with a resolution $(R) = 1$. Note that, although not indicated, the distance (Δd) between peak maxima may also be given in terms of time units and that Δd no matter how expressed, is independent of the void volume (V_0) of a column. (Reproduced from P. B. Hamilton, *et al*, *Anal. Chem.*, **32**, 1785 (1960), by permission of the authors and the American Chemical Society; reproduction was modified for use here.)

of about 2% (8, 10), as seen in Fig. 5-7. At $R < 1$ the 2σ band widths overlap and elution zones become more and more cross-contaminated, while at $R > 1$ the 2σ band widths do not touch, in which case the elution zones are completely separated. Separation is usually considered unsatisfactory when R is less than 0.8 (10).

As is implicit in Eq. 5-9, the separation of two components can be achieved either by decreasing the width of their elution curves or by increasing the difference between their peak maxima, or by having both of these occurring concurrently. This is illustrated in Fig. 5-8. Zone width is influenced by operating conditions such as volume of sample, flow rate, temperature, particle size of exchanger, composition of elution solutions, column dimensions and how well a column is packed. Optimizing these conditions *narrows* the peak width, or in other words, the closer to "true" equilibrium operating

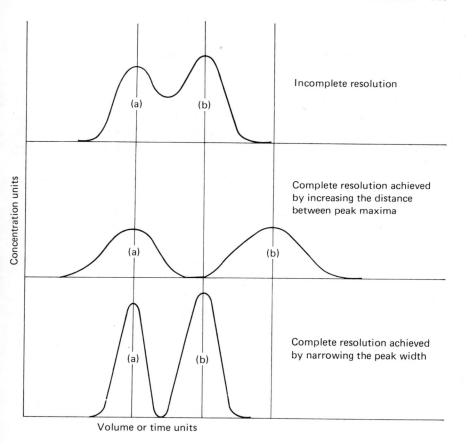

Fig. 5-8. Elution curves illustrating the achievement of good resolution between two components, (a) and (b), by either decreasing the width of their elution curves or by increasing the difference between their peak maxima.

conditions the smaller the band width of the elution cruve. The parameters affecting the shape of an elution curve can be considered jointly as a single quantity called *column efficiency* (8, 9, 10, 11), which is expressed, quantitatively, in terms of the number of theoretical plates (P) found for a column. From $\bar{v}$ and σ (or W), P is obtained from an elution curve (see Fig. 5-7) by the equation

$$P = \left(\frac{\bar{v}}{\sigma}\right)^2 = 16\left(\frac{\bar{v}}{W}\right)^2 \tag{5-10}$$

Since P is proportional to column length, a more useful measure of column efficiency, called *height equivalent of a theoretical plate* (HETP) or *plate height*, is obtained when the total number of plates is related to the column length (L) as given by (8, 14)

$$\text{HETP} = \frac{L}{P} \tag{5-11}$$

Hence, column efficiencies for columns of different lengths may be compared. The smaller the value for HETP, the greater will be the column efficiency.

If, after optimizing all the parameters that would narrow peak width, that is, give the smallest HETP, the separation of two components is still not satisfactory in terms of a desired resolution factor, the only recourse for improving resolution is to increase the length of the column. Before undertaking this operation, it must be determined that the exchanger shows some degree of selectivity, i.e., Δd of Eq. 5-9 is not equal to zero. Since, from the plate theory, Δd is proportional to the length (L) of a column and zone width (W) is proportional to the $\sqrt{L}$, then Eq. 5-9 can be written as

$$R = \frac{2 \, \Delta d}{W_a + W_b} \propto \frac{L}{\sqrt{L}} \propto \sqrt{L} \tag{5-12}$$

Thus, if the exchanger shows *some* degree of selectivity, no matter how small, it is theoretically possible to make a complete separation of two components given a long enough column. Consequently, given the resolution R' for two components separated unsatisfactorily on a column of length L_1, the length of column L_2 needed for a satisfactory separation having a desired resolution of R'' can be obtained by utilizing Eq. 5-12 to give

$$L_2 = \left(\frac{R''}{R'}\right)^2 \cdot L_1 \tag{5-13}$$

In considering an increase of column length, one must also consider that the time of analysis is also proportional to column length. Therefore any extra time spent for an analysis may not be worth the effort, since it is clear from Eq. 5-12 that resolution is not *directly* proportional to L. Thus even though increasing the column length by a factor of four quadruples the distance between peak maxima, the resolution increases by only a factor of two. Also, since solute bands spread as they travel the length of a column, they may be too dilute to detect after emerging from a "very long" column. Therefore even though a separation is theoretically possible, it may not be feasible on the basis of practical considerations.

REFERENCES

1. Baumann, F., in *Basic Liquid Chromatography*, an in-house publication of Varian Aerograph, Walnut Creek, Calif., 1971, Chap. 8; see also *Varian Aerograph Catalog* (*Instruments and Accessories*) issued in 1971.

2. Anderson, N. G., *Anal. Biochem.*, **4**, 269 (1962).

3. Spackman, D. H., Stein, W. H., and Moore, S., *Anal. Chem.*, **30**, 1190 (1958).

4. Hamilton, P. B., *Anal. Chem.*, **35**, 2055 (1963).

5. *Techniques in Amino Acid Analysis*, Technicon Instruments Company, Ltd., Chertsey, Surrey, England, 1966, Chap. III, p. 141; cited publication also is obtainable from Technicon Instruments Company, Tarrytown, N. Y.

6. Uziel, M., Koh, C. K., and Cohn, W. E., *Anal. Biochem.*, **25**, 77 (1968).

7. Singhal, R. P., and Cohn, W. E., *Anal. Biochem.*, **45**, 585, 1972.

8. Hamilton, P. B., Bogue, D. C., and Anderson, R. A., *Anal. Chem.*, **32**, 1782 (1960); Hamilton, P. B., in *Advances in Chromatography*, Marcel Dekker, Inc., New York, 1966, Vol. 2 (ed. by J. C. Giddings and R. A. Keller), Chap. 1.

9. Baumann, F., in *Basic Liquid Chromatography*, an in-house publication of Varian Aerograph, Walnut Creek, Calif., 1971, Chap. 3.

10. Karger, B. L., in *Modern Practice of Liquid Chromatography*, Wiley-Interscience, Inc., New York, 1971 (ed. by J. J. Kirkland), Chap. 1.

11. *Liquid Chromatography, Lab. Notes* 3, an in-house publication of Chromatronix Incorporated, Berkeley, Calif., 1971.

12. International Union of Pure and Applied Chemistry, Division of Anal. Chem., *Pure and Applied Chem.*, **8**, 553 (1964).

13. Algelt, K. H., in *Advances in Chromatography*, Marcel Dekker, Inc., New York, 1968, Vol. 7 (ed. by J. C. Giddings and R. A. Keller), Chap. 1.

14. Glueckauf, E., *Trans. Faraday Soc.*, **51**, 34 (1955).

Simplification of Some Common
Analytical Chemical Operations
SIX

A. NONCHROMATOGRAPHIC OPERATIONS

Many of the routine manipulations encountered in either qualitative or quantitative chemical procedures are often simplified and/or improved by making use of ion-exchange materials. For the sake of convenience and efficiency, the exchangers are usually contained in a column for these analytical operations. Thus, they are sometimes classified as chromatographic methods. However, since many of these operations are essentially batch processes or involve only simple separations, such procedures are more properly classified as *nonchromatographic operations* (1). In contrast, when on the order of hundreds of theoretical plates are needed to effect a separation (e.g., among several similar ionic species) the process is truly termed chromatography. An example of the nonchromatographic type of operation is the removal of a trace component from a solution by passing it through an exchanger bed. An example of ion-exchange chromatography is the separation of a mixture of amino acids. Nonchromatographic applications are treated in this chapter, while ion-exchange chromatography and related methods are discussed in the chapters that follow. In each case, of many hundreds, only a few typical examples are presented.

The resinous exchangers are almost always employed in nonchromatographic operations. Therefore these types of exchangers are the materials cited in the examples of this chapter. If not applicable for a particular operation, other exchange materials may be substituted (see Chapter 3) for the resinous exchangers.

126

B. CONVERSION OF ONE COMPOUND TO ANOTHER

This is a common laboratory operation wherein a needed compound is prepared from chemicals that are at hand. For instance, a conversion of CsCl to $CsNO_3$ occurs when a solution of CsCl is rinsed through a fixed bed of an anion exchanger in the nitrate form in accordance with the ion-exchange reaction:

$$CsCl + RNO_3 \rightleftharpoons CsNO_3 + RCl \qquad (R = \text{an exchange site of the resin})$$

$$(6\text{-}1)$$

Likewise NH_4Br may be prepared from NaBr by allowing a solution of NaBr to percolate through a cation exchanger in the ammonium form as given by the reaction:

$$NaBr + RNH_4 \rightleftharpoons NH_4Br + RNa \qquad (6\text{-}2)$$

In addition to the preparation of salts, a variety of acids or bases may by synthesized by this technique. The net result of this type of conversion is that one ion of an electrolyte is removed from solution and is replaced by another without the introduction of an additional soluble electrolyte. In other words, the co-ion present in the solution passed through the column appears in the effluent with the displaced counter-ion that was initially bound to the exchanger, and thus the desired compound is formed.

The ease of conversion is dependent upon the relative exchange affinities (the ratio of their distribution coefficient) of the two counter-ions in question. For example, if potassium chloride is to be converted to potassium acetate by this technique, little more than the required theoretical amount of exchanger in the acetate form is needed in the column. That is, the breakthrough and the theoretical capacities are essentially identical for this particular reaction, since the affinity or selectivity coefficient of the chloride ion is much greater than that of the acetate ion. If the reverse conversion, acetate to chloride, is attempted, a large excess of anion exchanger in the Cl-form is needed in the column in order to avoid "leakage" of the acetate ions into the effluent. In this case, the breakthrough capacity is only a small fraction of the theoretical capacity of the column. Thus, before attempting a bulk synthesis by this technique, a few pilot runs (e.g., using a small quantity of resin contained in medicine droppers or pipettes) should be carried out to establish the quantity of exchanger needed for a quantitative conversion. Simple tests on the effluents are all that are generally needed to establish the experimental conditions. This type of operation is adaptable to almost any scale, the limiting factor being the size of equipment that is available.

C. THE STANDARDIZATION OF A SALT SOLUTION BY ION EXCHANGE

The normality of salt solutions may be determined accurately and rapidly by ion exchange. This technique is used routinely to determine the total salt concentration of unknowns as well as to determine more precisely the concentration of solutions whose normality is known only approximately (e.g., solutions prepared from hygroscopic salts or by the "rough" weighing of a salt).

The method consists merely of rinsing a sample through a column of strong-acid cation exchanger in the H-form and titrating with standard alkali the released acid that appears in the effluent. Any acidity in the original sample must be determined by direct titration of the sample prior to the ion-exchange treatment and subtracted from the result.

The following is a general procedure that may be used to standardize a salt solution. Consider that the solution to be checked is approximately 1 molar.

1. Slurry 3–5 ml of a strongly acidic cation exchanger (200–400 mesh, 4–8% crosslinked) in the H-form into a 1.0 to 1.5 cm diameter column of short height (3–5 cm high).

2. After the exchanger bed is thoroughly washed with deionized water, place a 1-ml aliquot of the salt solution on the exchanger bed.

3. Wash the salt sample through the column with several portions of water.

4. Titrate the effluent (including the water rinses) with 0.1000 N* standard alkali using methyl red or phenolphthalein as the indicator.

5. Calculate the normality (N) of the salt solution from

$$N = \text{ml of standard alkali} \times 0.1000 \qquad (6\text{-}3)$$

D. REMOVAL OF INTERFERING IONS

In many instances anions or cations that interfere with the analysis for a particular ion or constituent are conveniently removed by an ion-exchange step prior to the actual determination. Interferences may be caused by complex-ion formation, co-precipitation, the presence of an undesired colored species, the interaction of foreign ions with reagents added for a spectro-

---

*The concentration of standard alkali used for the titration may be varied to accommodate the concentration of the salt solution. For instance, if a 0.1 N salt solution is to be checked, then 0.0100 M alkali is a suitable concentration for the standard alkali.

photometric analysis, and so on. For instance in the gravimetric determination of sulfate, sodium, aluminum, ammonium, and ferric ions are co-precipitated with the barium sulfate. This gives an erroneous value for the percentage of sulfate in a sample. To avoid such errors prior to the addition of barium ions, the sample is simply passed through a column of a strong-acid cation exchanger in the H-form. By this step all cations initially present in the sample are replaced with hydrogen ions; the net result is a sulfuric acid solution. In a similar fashion, cations such as calcium, iron, and aluminum are removed by cation exchange prior to the determination of phosphorus by the precipitation of magnesium ammonium phosphate. Another example of this kind of application is found in the removal of IO_3^- and IO_4^- ions following a periodate oxidation of carbohydrate material. These ions interfere with the colorimetric determination of CH_2O and other treatments that are sometimes carried out on periodate-oxidized samples. Passage of the sample through an anion exchanger in the acetate form removes the IO_3^- and IO_4^- ions (see Section G). In a very pertinent application, Kempf (2) utilized a strong-acid exchanger to sorb interfering cations prior to the determination of fluoride in potable drinking water. Fluoride was determined directly in the column eluate.

Sometimes the constituent sought is isolated on a column while interfering substances are not retained and are washed through the exchanger bed, thus effecting a removal of undesired components. Shrimal (3) used this technique for the determination of silver in the presence of mercury. In such mixtures, high concentrations of mercury interfere in the determination of silver by the Volhard method. Shrimal solved this problem by isolating silver in the presence of EDTA on a 1.0×15 cm column of Dowex 50W-X8 in the sodium form. Mercury(II) forms a stable anionic complex with EDTA at pH 4–5 and passes through the cation column unchanged. The silver-EDTA complex is not stable and is dissociated on the column, resulting in the retention of silver on the exchanger bed. After washing out the mercury complex, the silver is eluted with 250 ml of 16% nitric acid and titrated with standard ammonium thiocyanate.

E. PREPARATION OF DEIONIZED WATER

The production of high purity water approaching theoretical resistivity can result when aqueous solutions are demineralized by ion-exchange treatment. Rather than use a two-step process (see Fig. 6-1) in which a salt solution is passed successively through a strong-acid exchanger in the H-form and a strong-base anion exchanger in the OH-form, it is more convenient to use a mixed-bed exchange cartridge (these are available commercially, both exchangers being present in a single package) that may be hooked

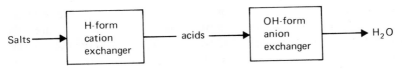

Fig. 6-1. Dual column demineralization of salt solutions.

directly to a tap water supply. A resistance meter is usually a part of such an apparatus, and when the cartridge is used properly, water with a specific resistance of 250,000 to 1,000,000 ohm cm, or as little as 2 ppm (parts per million) of total electrolyte, is easily obtained (4). These units are expendable and when they become exhausted, the old cartridge is simply replaced.

Ion exchange removes completely all electrolytes, including weakly ionized compounds such as silicic, carbonic, and boric acids and phenol. Nonelectrolytes are not removed and they remain in the demineralized water. The exchangers themselves may be a source of nonionic impurities, especially when flow through a cartridge is interrupted. At start-up time, material arising from the slight dissolution of the exchanger bed may appear in the effluent. Thus, collection of water should not be attempted until thorough rinsing of the exchanger bed has taken place.

Mixed monobed ion-exchange units are particularly effective where such units are attached to an ordinary distilled water supply. This allows the preparation of essentially solute-free water.

F. PURIFICATION OF ORGANIC COMPOUNDS

There are three ways in which exchange materials may be employed for the purification of organic compounds. The ionic impurities in solutions of neutral substances such as sugars, glycerol, ethylene glycol, lower molecular-weight aliphatic alcohols, ketones, etc. can be removed in the same manner that water is deionized (see Section E). Solutions of the organic compounds are merely passed through a combination of exchangers, one a cation-exchange resin, the other an anion-exchange resin. In some cases, even batch-stirring processes can suffice. Secondly, if the compounds themselves are ionizable, they can interact with exchange materials. In this manner, organic acids, bases, or salts may be sorbed by the appropriate exchange material, thus making possible their isolation and recovery from nonionic organic compounds as well as from inorganic substances. The size and type of the organic molecule is a critical factor in this type of purification. Therefor the precautions noted in Chapter 3 regarding the selection of the proper exchanger should be evaluated before attempting organic purifications. Thirdly, some organic compounds can undergo reactions that lead to the formation of addition compounds or complexes that are ionized, and hence in this form can enter into exchange reactions. Examples of this type of species are given by the bisulfite addition complexes of aldehydes or ketones (5) and the borate complexes of polyhydroxy substances (6). Another method of purification in this area is that of ion exclusion, which is treated separately in Chapter 8.

The utilization of ion exchangers in the field of organic chemistry is reviewed by Walton (7). Much of the work done in this area after Walton's review is summarized in the annual reviews (fundamental) published on alternate years by *Analytical Chemistry*.

G. DETERMINATION OF IODATE FOLLOWING PERIODATE OXIDATION OF α-GLYCOL GROUPS

The extent of α-glycol cleavage in a polyhydroxy substance corresponds stoichiometrically to the amount of iodate (IO_3^-) found in a mixture after a carbohydrate compound has been oxidized with periodate (IO_4^-). Most methods of determining IO_3^- in periodate-oxidized samples are based upon the back titration of excess IO_4^- that remains following an oxidation reaction. However, the amount of IO_3^- may be determined by a direct spectrophotometric measurement of IO_3^- free of interfering IO_4^- by ion-exchange analysis.

The ion-exchange method is summarized in the following five essential steps: (1) the carbohydrate sample is oxidized with IO_4^-; (2) the sample is passed through a small anion exchanger that sorbs both IO_3^- and IO_4^-; (3) IO_3^- is selectively eluted from the exchanger with ammonium chloride solution (IO_4^- remains sorbed); (4) the absorbance of the eluted IO_3^- is determined at 232 nm; (5) and the absorbance value is converted into moles of IO_3^-, which is equal to the moles of α-glycol oxidized.

One advantage of the ion-exchange method is that it removes quantitatively both IO_3^- and IO_4^- from the oxidized sample, which passes through the exchanger (the eluate of step 2). This allows further direct treatment of periodate-oxidized carbohydrates, such as those involved in structural synthetic or preparative work. Also, tests for the presence of formaldehyde (arising from a primary alcohol group contained in the original carbohydrate) can be done directly on the eluate of step 2. A disadvantage of this method is that it is not applicable to those polyhydroxy substances (e.g., aromatic derivatives or acidic materials) that sorb to the resin (step 2) and that, upon subsequent elution (step 3), could interfere with the spectrophotometric measurement (step 4).

Different sample conditions (e.g., pH or the presence of salts or buffers) may alter somewhat the general procedure described in the section that follows. More details and variations of this method are described elsewhere by Khym (8).

Procedure   The compound (25–150 μmoles) is oxidized in 10 ml of 0.02 M sodium periodate. After a given time period, the sample is rinsed through a column (about 1 cm in diameter) containing about 3 ml of a strong-base polystyrene anion exchanger in the acetate form. Following a 25-ml water wash, IO_3^- is selectively eluted from the column with 100 ml of 0.1 M ammonium chloride solution. The absorbance of eluted IO_3^- is read in a quartz absorption cell at 232 nm. The extent of α-glycol cleavage is calculated from the following equation:

μmoles IO_3^- found $= \mu$moles of α-glycol oxidized

$$= \frac{\text{total absorbance of NH}_4\text{Cl} - \text{solution absorbance of blank*}}{E_{\text{mM}}^{IO_3^-}} \qquad (6\text{-}4)$$

where $E_{\text{mM}}^{IO_3^-}$ is the millimolar extinction coefficient for IO_3^-. The value for this coefficient is approximately 0.90 and is obtained by adding 200 μl of 0.1 M (100 mM) sodium periodate to 200 μl of 1 M glycerol in a 50-ml volumetric flask. After diluting to volume with water the absorbance (A) of the reduced periodate solution is determined, as before, at 232 nm. The E_{mM} is calculated from the equation:

$$E_{\text{mM}}^{IO_3^-} = \frac{A_{232}^{IO_3^-} \times 250}{100 \text{ mM}} \qquad (6\text{-}5)$$

where $A_{232}^{IO_3^-}$ is the absorbance of the iodate solution and 250 is the dilution factor. The extinction coefficient of IO_3^- also can be determined directly by making the appropriate measurements of a weighed quantity of reagent grade sodium iodate dissolved in water.

H. CATALYSIS

A strong cation exchanger in the H-form or a strong anion exchanger in the OH-form may be considered, respectively, as a solid strong acid or base. As such, these exchangers can behave as catalysts in certain aqueous reactions, if the solutes present can penetrate the exchanger matrix and undergo reaction with the active counter-ions (either H^+ or OH^-) of the exchange material. This process is usually carried out as a batch procedure, although in some instances the material to be reacted is passed through a column. The chief advantage of the column method is that the reaction

---

*Blank determination gives rise to an absorbance of about 0.05 due to the presence of acetate absorption at 232 nm.

products are easily separated from the exchangers. In the batch method, the products are recovered by filtration, decantation, or centrifugation, and in the column method the products appear in the effluent by the proper rinsing of the exchanger bed.

Khym and Cohn (9) used resin catalysis to establish the position of the phosphate groups in the ribose moiety of the isomeric adenylic acids (e.g., adenosine 2′-, 3′-, and 5′-phosphates). The N-glycosyl linkages of the individual nucleotides were hydrolyzed with a polystyrene sulfonic acid exchanger. In short time periods (e.g., about 30 sec), the products of the reaction are adenine (in each case) and the daughter ribose phosphate originating from the parent nucleotide, that is, adenosine 2′-phosphate yields ribose 2-phosphate, and so on. The ribose phosphates were identified as their borate complexes by ion-exchange chromatography by a procedure that separates all the isomeric ribose phosphates (see Chapter 7). The important feature of this method is that the anionic ribose phosphates, once formed, are rejected from the environment of the high acidity of the cation exchanger into a water solution of their own pH where only minimal isomerization occurs. This eliminated a problem of the older hydrolytic experiments, using soluble acids like HCl, in which phosphomigration occurred in the ribose moiety during the hydrolysis period. Another advantage is that adenine remains sorbed to the exchanger and can be removed with base for material balance calculations after separation of the resin and aqueous phases. The reaction is carried out by simply adding nucleotide (solid) to a mechanically stirred slurry of resin and water maintained at a temperature of 100°C. The ratio of water (ml) to wet Dowex 50 (ml) to nucleotide (g) was 1:1:0.01 for analytical investigations. After short periods of time (a few minutes or less) the reaction mixture was cooled in an ice bath and then the phases were separated by filtration.

Noggle (10) employed cation resin catalysis to hydrolyze the O-glycosidic bond of sucrose. The disaccharide, 5 to 250 mg dissolved in 5 ml of water, was mixed with 1 g of moist acid exchanger (several commercial sources were used interchangeably) in a test tube that was immersed in a boiling water bath for about 40 min. After the reaction, the hydrolytic products, D-fructose and D-glucose, were recovered in the aqueous phase by filtration.

Qureshi, *et al.* (11) developed a column method of catalysis for the hydrolysis of water-soluble aliphatic amides and esters. The amide or ester, dissolved in about 10 ml of water, is passed through a 30 × 0.8 cm column of IR-120 in the H-form at a flow rate of 2 ml per min at 80°C. The effluent is recycled three times and finally the organic acids arising from the hydrolysis are titrated with standard alkali.

Resin catalysis has been used for esterification, alkylation, epoxidation and for the hydrolysis of nitrites, proteins, and other organic products. A detailed review on this subject has been compiled by Astle (12). For reports of resin catalysis published after Astle's summary, see the reviews on ion exchange that appear in *Analytical Chemistry* on alternate years.

A recent new development in this field is the permanent fixation of enzymes to polymeric supports, such as to the matrix material of ion exchangers. In a sense this may be considered an extension of exchanger catalysis. The main interest in water-soluble enzymes immobilized on a water-insoluble polymer stems from their possible use as heterogeneous specific catalysts in research and industry. The advantages of working with "solid enzymes" are essentially the same as those listed previously for strong acid or base resin catalysis. In biochemical work, there is a great advantage in the convenience of recovering the product of an enzyme reaction simply by filtration, decantation, or centrifugation, and thus having it easily ready for reuse.

Enzymes have been immobilized on the ion-exchange resins (the divinylbenzene type) but the cellulose exchangers have been more extensively utilized. Water-insoluble enzymes have been prepared for the study of biologically active proteins, nucleic acids, polysaccharides, and a variety of lower molecular-weight substances. There are literally thousands of applications of this type of catalysis. Stark (13) has written a comprehensive review on this subject. Goldstein (14) has described water-insoluble proteolytic enzymes affixed to various polymeric materials. A compendium of references on the immobilization of enzymes can be obtained from the Corning Biological Products Group (Medfield, Mass. 02052). *Annual Reviews of Biochemistry* is another literature source dealing with this subject.

I. RECOVERY OF TRACE CONSTITUENTS

Ion exchange is most suitable for the task of concentrating the trace constituents of a solution. The procedure consists simply of passing large volumes of very dilute solutions through a relatively small ion-exchange column. Once collected on the exchanger, the trace ions are recovered in a small volume upon elution of the column with an appropriate eluent. Volume reductions of several orders of magnitude can be easily achieved by this method. It is not uncommon for solutes to be concentrated 100- to 1000-fold so that, if not applicable before the concentration step due to sensitivity limitations, conventional analytical procedures can be used to quantitatively determine the now-concentrated trace components. Ideally, if the trace components are to be recovered quantitatively, large amounts of other electrolytes should be absent. However, in some cases, the affinity of a minor component in solution is so much greater than that of a major constituent that the collection of the trace substance still occurs.

Applications of this technique, both on a large and small scale, are many and are varied. Due to more stringent antipollution regulations, the ion-exchange method for collecting trace constituents has become increasingly important for monitoring the effluents of industry. In more general environmental studies, the technique is almost ideal for the analysis of streams, rivers, lakes, or ocean waters. The trace ionic components of foods and beverages are often collected on ion-exchange materials, either for the purpose of concentrating them prior to analyses, or for altering the composition of the food product. Many valuable metals present at trace concentrations but in large volumes from electroplating processes are recovered by unit ion-exchange processes, some operations even resulting in economic gain. On a laboratory scale, this technique is often used to recover components separated by chromatographic procedures. Such components are usually present in eluates in very low concentrations compared to the concentration of the electrolyte that was employed to elute them. In these cases, often a simple adjustment in pH will change the net charge of a molecule so that at the new pH its affinity for a given exchanger is much greater than that of the eluting solution. Hence, the eluted species may be concentrated on a much smaller column than that from which it was isolated. By elution of the second column with a volatile eluent, the component can often be recovered in a pure state.

1. Concentration of Trace Substances from Water Supplies

One advantage of the ion-exchange method is that large volumes of water can be passed through the exchanger at the test site, hence the transportion of bulky samples to the laboratory is avoided. Only the column must be brought back; the constituents are analyzed upon eluting them from the exchanger.

Prior to the photometric determination of fluoride in potable waters, Kelso *et al.* (15) concentrated this halogen on the acetate-chloride form of Dowex 2. Water samples (~ 50 ml) were passed through about 12 ml of exchanger contained in a 30×1 cm column at 3 ml per min. Fluoride was eluted as its beryllium complex (BeF_4^-) with dilute beryllium acetate solution. By this treatment, interfering substances were eliminated and fluoride was determined directly in the eluate by a photometric procedure.

Szabo and Joensuu (16) determined the content of barium, strontium, and calcium in sea water after these alkaline earths were concentrated by an ion-exchange step. In this procedure, 100-ml samples of sea water were rinsed through a column (1.5×13.5 cm) of Dowex 50-X8 (100–200 mesh) in the H-form at a flow rate of 0.6 ml per min. The column was eluted with 1 M HCl to remove the alkali metals and the major portion of magnesium. When calcium appeared in the effluent (ascertained by an indicator method), the concentration of HCl was increased to 5 M, which elutes calcium, barium, and strontium. After evaporation of the HCl, the alkaline earths were precipitated as their fluorides and determined by emission spectrographic analysis.

Ordinary cation and anion exchangers are not as efficient as the chelating resins for the analytical concentration of trace elements from sea water. The chelating resins have a crosslinked polystyrene matrix to which iminodiacetic acid groups [$—CH_2N(CH_2COOH)_2$] are attached. Thus the exchanger is extremely selective for divalent metals ions, especially the transition elements. Riley and Taylor (17) used 6.0 × 1.0 cm columns of Chelex 100 or Permutit S1005 to remove various trace elements from liter quantities of sea water. The trace constituents were determined by atomic absorption spectrophotometry after stripping them from the chelate resins with 2 N acids or 4 M ammonia. As seen in the data summarized in Table 6-1, many trace elements, both cations and anions, are recovered quantitatively by this process. Siegel and Degens (18), see also (19), isolated trace amounts of amino acids from sea water using a chelating exchanger in the cupricammonium form. A metal such as copper when bound to a chelate exchanger still retains its ability to coordinate ligands. One ligand may be replaced by another in a process called ligand exchange (see Chapter 9). Since the exchanger has such a strong affinity for a complexing metal such as copper, ion exchange does not occur, at alkaline pH, when a strong electrolyte like sea water is passed through a bed of the chelate exchanger in the cupricammonium form. However, ligand exchange can occur and ammonia may be replaced by amino acids present in sea water. About 10 ml of the copper-loaded resin packed in narrow tubes will isolate the microgram amounts of amino acids present in 2-l volumes of sea water. Concentrated ammonia (3.0 M) is used to free the sorbed amino acids which are then analyzed by a sensitive chromatographic procedure.

A polyamine-polyurea resin synthesized by Dingman *et al.* (20) was utilized to concentrate metals such as Cu, Ni, Zn, and Co in concentrations as low as 4 parts in 10^{10}. High concentrations of alkali and alkaline earth metals are not complexed by the synthesized exchanger and they do not compete with the complexation of trace heavy metals. Thus, the polyamine-polyurea resin should be particularly important for the analysis of trace heavy metals in natural waters (both sea and fresh water).

Dollman (21) determined sulfate in natural waters by a titrimetric method after passage of water samples containing about 5 mg of sulfate and of total ionic strength not greater than 0.02 N through a bed (about 20 ml of resin) of strong-acid exchanger in the H-form. The effluent is evaporated under controlled conditions which completely volatilizes ordinary acids while quantitatively retaining the sulfuric acid. Titration with standard base

TABLE 6-1

Data on Adsorption and Elution of Trace Elements from Sea Water with Chelex-100 (50–100 mesh) and 20 ml of Eluting Agent

	pH for adsorption	Retention (%)	Eluant	Total recovery (%)
Aluminium	7.6	0	—	0
Arsenic (AsO$_4$$^{3-}$)	7.6	0	—	0
Barium	5.0[a]	25	2 N HNO$_3$	25
Bismuth	9.0[a]	100	2 N HClO$_4$	100
Cadmium	7.6	100	2 N HNO$_3$	100
Caesium	7.6	0	—	0
Cerium (Ce^{3+})	9.0[a]	100	2 N HNO$_3$	100
Chromium (Cr^{3+})	5.0[a]	25	2 N HNO$_3$	10[b]
Cobalt	7.6	100	2 N HCl	100
Copper	7.6	100	2 N HNO$_3$	100
Indium	9.0[a]	100	2 N HNO$_3$	100
Lead	7.6	100	2 N HNO$_3$	100
Manganese	9.0[a]	100	2 N HNO$_3$	100
Mercury (Hg^{2+})	7.6	85	2 N HNO$_3$	40[b]
Molybdenum (MoO$_4$$^{3-}$)	5.0[a]	100	4 N NH$_4$OH	100
Nickel	7.6	100	2 N HNO$_3$	100
Phosphorus (PO$_4$$^{3-}$)	7.6	0	—	0
Rhenium (ReO$_4$$^-$)	7.6	90	4 N NH$_4$OH	90
Scandium	7.6	100	2 N HNO$_3$	100
Selenium (SeO$_4$$^{2-}$)	7.6	0	—	0
Silver	7.6	100	2 N HNO$_3$	90[b]
Thallium (Tl$^+$)	7.6[a]	50	2 N HNO$_3$	50
Thorium	7.6	100	2 N H$_2$SO$_4$	100
Tin (Sn$_4$$^+$)	7.6	85	2 N HNO$_3$	60[b]
Tungsten (WO$_4$$^{2-}$)	6.0[a]	100	4 N NH$_4$OH	100
Uranium (UO$_2$$^{2+}$)	7.6	0	—	0
Vanadium (VO$_3$$^-$)	6.0[a]	100	4 N NH$_4$OH	100
Yttrium	9.0[a]	100	2 N HNO$_3$	100
Zinc	7.6	100	2 N HNO$_3$	100

[a]Optimum pH value.
[b]Maximum percentage removable from resin.
Source: Reproduced from J. P. Riley and D. Taylor, *Anal. Chim. Acta*, **40**, 479 (1968), by permission of the authors and Elsevier Publishing Company.

completes the determination. The method gives excellent agreement with standard volumetric and gravimetric procedures for determining sulfate. With prior sample dilution or concentration, sulfate in parts per billion to parts per hundred may be determined by this method.

2. Recovery of Trace Constituents in General Environmental Work

In public health work, Porter *et al.* (22) determined strontium-90 in milk samples by isolating the yttrium-90 daughter as its anionic citrate complex on Dowex 1-X8. In this method stable yttrium carrier is added to a 1-l sample of milk which is then passed successively through the Na-form of Dowex 50W (140 ml of resin in a 5 cm diameter column) to remove alkaline earth ions, and then through the Cl-form of Dowex 1-X8 (30 ml of resin in a 3 cm diameter column) which retains the yttrium, probably as its citrate complex. The yttrium is eluted with HCl, then precipitated as its oxalate prior to radiochemical analysis. A more general procedure (23, 24) for determining stable strontium or radiostrontium in environmental samples includes an ion-exchange step in which EDTA is used to separate strontium from other alkaline earths. In this procedure an alkaline carbonate fusion of an ashed sample followed by dissolution in dilute acid gives a preliminary separation. Enough EDTA is added to complex about 80% of the calcium, and after adjusting the pH to 5.1, the sample (now contained in about a 500-ml volume) is passed over the Na-form of Dowex 50W-X8 (20 ml of resin in a 1.9 cm diameter column) at 60-70 ml per min. Under these conditions the exchanger has such a strong affinity for strontium that dilute EDTA solution (2%) and dilute acid washes (0.5 M HCl), respectively, will remove any other residual alkaline earth ions as well as the alkali metal ions. After elution with 3 M HCl, the strontium is concentrated by evaporation and is determined by flame photometry or by radiochemical analysis.

Talvitie (25) has developed an anion-exchange method to determine plutonium in environmental and biological samples such as urine, animal tissue, bone, soil, fresh waters, or the ocean. After pretreatment of the samples (either wet or dry ashed) Pu(IV) is stabilized with hydrogen peroxide and is sorbed as its negative chloro complex on Dowex 1-X2 (20 ml of settled resin in a 1.45 cm $\times$ 25 cm column) from 9 M HCl solution. Co-sorbed iron is removed with 7.2 M HNO$_3$, then Pu is reductively eluted with 1.2 M HCl-0.6% H$_2$O$_2$ and electroplated from 1 M (NH$_4$)$_2$SO$_4$ at pH 2 for alpha spectrometric determination.

Ion exchangers have found extensive use in the atomic energy field to process uranium fission-product solutions. Collections of papers on this important problem of the disposal of radioactive liquid wastes may be found in the *Peaceful Uses of Atomic Energy* series that are published by the United Nations.

3. The Isolation of Trace Elements in Food Products

In the food industry, ion exchangers are used to isolate trace substances for analysis and also to simply remove trace materials in bulk processing schemes. For instance in the treatment of alcoholic beverages and fruit juices, ion exchangers are utilized to remove traces of heavy metals (e.g., Cu, Fe, Cr) as well as to control the acidity of these liquids. Exchangers are used in a similar manner by the dairy industry to modify and improve the quality of milk products. Cranston and Thompson (26) were the first to apply the technique of concentrating a trace substance by ion exchange for the purpose of analyzing food products. They used a strong-acid exchanger to concentrate copper, which may be present in milk in concentrations of less than 1.0 ppm. The copper was eluted with concentrated HCl and was determined by standard analytical procedures. Analyses of this type are described in reviews by Smith (27) and by Austerweil (28). Also many chromatographic methods are described in the reviews on the analysis of food products in the application section that appears on alternate years in *Analytical Chemistry*.

4. Separation of Trace Constituents in the Ore and Metal Industries

The ability of exchangers to concentrate selectively many valuable ions present in solution in such small amounts that they cannot be recovered by precipitation or evaporation procedures makes ion exchange a potent tool for attacking the waste problems of the metal industries. Trace collections of such valuable metals as gold, silver, zinc, copper, cobalt, manganese, nickel, and chromium are carried out in hydrometallurgical operations utilizing both anion and cation exchangers. Ion exchange is also used for analytical purposes to concentrate trace elements in rocks and ores. A few examples of these types of applications are given.

Thompson and Miller (29) give several examples of how ion-exchange unit processes are employed in the metal finishing industries to comply with the more strict antipollution regulations and in doing so also make attractive economic gains. For example, chromic acid baths containing 250 to 400 g per l of chromic acid are freed from such impurities as iron, copper, zinc, and trivalent chromium that are accumulated in plating processes. The contaminated chromic acid solution is merely passed through a cation-exchange column operated in the hydrogen cycle, whereby the metal-ion impurities are exchanged for hydrogen ion and in the case of those combined with chromates, chromic acid is reformed. Deionized water used for rinsing plated articles can be recovered for reuse by passing the rinse waters through a two-bed ion-exchange system. The first unit contains a hydrogen form cation exchanger that removes positively charged ions and the second vessel contains an anion exchanger operated in the hydroxide form that neutralizes the acid released from the cation exchanger. Precious metals such as gold and platinum are recovered from plating solutions as their anion radicals on strong-base anion exchangers.

Biechler (30) demonstrated that trace copper, lead, zinc, cadmium, nickel, and iron in industrial waste waters may be concentrated on a chelating resin. In this process contaminated waters are buffered to pH about 5.2 and passed through columns of Dowex A-1. The separated metals are stripped with 8.0 M nitric acid.

High concentrations of alkali salts interfere in the determination of trace constituents in ores by atomic absorption after a basic fusion and solution of the sample in acid. Freudiger and Kenner (31) eliminated this interference in the determination of cobalt, copper, manganese, zinc, nickel, iron, and lead in ore samples. The fused material is dissolved in acid, and the pH adjusted to 7.0. The solution is then poured through a column of Chelex 100 (Dowex A-1) which retains the transition metals but not the alkali metals. The retained metals are stripped from the chelating resin with 3 N HCl and after diluting with water are measured by atomic absorption.

Pitts and Beamish (32) described a cation-exchange method that quantitatively separates gold from large excesses of base metals such as iron, copper, and nickel that are present in gold ores. In this procedure the base metals are sorbed to a cation exchanger from 0.1 M HCl solution, while gold as its tetrachloro anion passes through the column and is recovered by evaporation prior to its determination gravimetrically or colorimetrically. In this operation it is essential to maintain the pH of the influent between 0.8 and 1.5; otherwise the separation is not quantitative, some gold being partially sorbed to the cation exchanger.

Sellers (33) determined small quantities of lead in steel samples by isolating the lead as its anionic chloro complex on an anion exchanger. After dissolution, steel samples were taken-up in 1 N hydrochloric acid and applied to a 1.3 × 13 cm column of Dowex 1-X8 in the Cl-form. The chloro-lead complex was eluted with 12 N hydrochloric acid. After evaporation of the eluate, lead was determined by atomic absorption methods.

5. Recovery of Solutes from Chromatographic Peaks

Easily amenable to this type of recovery are such substances as amphoteric metals, ionic complexes, and organic molecules that have both acidic and basic functional groups (e.g., amino acids or nucleotides). In these classes of compounds, it is obvious that a change in pH can bring about a change in the ionic character or net charge on the species and hence change its affinity for a particular exchanger.

An illustration of this is seen in the recovery of the rare earths after their separation as citrate or tartrate complexes on cation exchangers in the pH range of 2 to 7, where the positive charge of these metals is partially neutralized due to complex formation. Once separated, the addition of acid destroys the citrate or tartrate complex and the uncomplexed metal increases its positive charge and exhibits a greater affinity for the same exchanger than before. Thus, the rare earths can be concentrated on a small column of that exchanger and eventually be recovered in a small volume of eluate. This same type of volume reduction procedure could apply to the recovery of the transition-ammonia complexes such as the Cu, Cd, Ni, or Zn species or the amphoteric metals such as Cr, Zn, or Al if they are to be recovered from eluates.

Ion-exchange volume reduction techniques are commonly used in the biochemical field to recover material isolated from column work. For example, following a preparative-scale separation of deoxynucleotides on an anion exchanger (12 cm × 33 sq. cm) at low pH, individual fractions (12 to 20 liter) were made alkaline (so as to fully ionize the phosphate group of the nucleotides) with ammonia and readsorbed on smaller columns (containing about 1 ml of anion exchanger per 130 mg of nucleotide to be adsorbed). Upon elution, the nucleotides were recovered in concentrations of the order of 20 mg of deoxynucleotide per ml of dilute chloride solution (34).

The exchange potential can be altered in another manner to favor sorption instead of elution, by replacing one ion in the eluate with another. In a preparative method for the separation of ribose phosphates (35), millimole amounts of product contained in 20-l fractions were recovered in 100 ml of a volatile solvent. In this procedure it was necessary to remove sulfate ions (a constituent of the eluent) and replace them with monovalent acetate ions. This altered the exchange potential so as to favor the sorption of the ribose phosphates on small acetate exchangers (containing about 12 times less exchanger than was used for the initial separation). The replacement was accomplished by adding barium acetate to precipitate the sulfate ions in each of the fractions to be concentrated. After the barium sulfate had settled, the clear supernatant was passed through the small volume reduction columns, which were subsequently eluted with dilute acetic acid to recover the ribose phosphates.

Concentration and desalting by ion exchange is also applicable to analytical systems where the identity of a constituent in a chromatographic peak may have to be checked by other means. Dreze *et al.* (36) recovered micro amounts of individual amino acids from the concentrated citrate and phosphate buffer solutions (also present are detergents and autooxidants) used as the eluents in the ion-exchange chromatographic systems. The separated peaks (1–4 ml) containing the neutral and acidic amino acids were sorbed onto small columns (2 × 0.9 cm) of Dowex 2 in the OH-form. Washed through the column were cations, polyoxyethylenelauryl alcohol (BRIJ-35) used as the detergent, and the thiodiglycol used as the oxidant. Elution with 1 *M* acetic acid removes the amino acids but leaves the stronger, nonvolatile citric and phosphoric acids of the buffers, as well as chloride ions, on the resin. The basic amino acids were sorbed onto the hydrogen form of Dowex 50 and eventually recovered by elution with 4 *N* HCl.

REFERENCES

1. Rieman III, W., and Walton, H. F., *Ion Exchangers in Analytical Chemistry*, Pergamon Press, New York, 1970, Chap. 5.

2. Kempf, T., *Z. Anal. Chem.*, **244**, 113 (1969).

3. Shrimal, R. L., *Talanta*, **18**, 1235 (1971).

4. Dean, J. A., *Chemical Separation Methods*, Van Nostrand Reinhold Company, New York, 1969, p. 108.

5. Gabrielson, G., and Samuelson, O., *Svensk. Kem. Tidskr.*, **62**, 214 (1950); *Acta Chem. Scand.*, **6**, 738 (1952); see also Samuelson, O., *Ion Exchangers in Analytical Chemistry*, John Wiley & Sons, Inc., New York, 1953, Chap XVI.

6. Khym, J. X., and Zill, L. P., *J. Am. Chem. Soc.*, **74**, 2090 (1952).

7. Walton, H. F., in *Ion Exchangers in Organic and Biochemistry*, Interscience Publishers, Inc., New York, 1957 (ed. by C. Calmon, and T.R.E. Kressman), Chap. 35.

8. Khym, J. X., in *Methods in Carbohydrate Chemistry*, Academic Press, Inc., New York, 1972 (ed. by R. L. Whistler and J. N. BeMiller), Article 12.; see also Khym, J. X., and Cohn, W. E., *J. Am. Chem. Soc.*, **82**, 6380 (1960).

9. Khym, J. X., and Cohn, W. E., *J. Am. Chem. Soc.*, **76**, 1818 (1954).

10. Noggle, G. R., *Plant Physiol.*, **28**, 736 (1953).

11. Qureshi, M., Qureshi, S., and Singhal, S. C., *Anal. Chem.*, **40**, 1781 (1968).

12. Astle, M. J., in *Ion Exchangers in Organic and Biochemistry*, Interscience Publishers, Inc., New York, 1957 (ed. by C. Calmon and T. R. E. Kressman), Chap. 36.

13. Stark, G. R., *Biochemical Aspects of Reactions on Solid Supports*, Academic Press, Inc., New York, 1971.

14. Goldstein, L., in *Methods in Enzymology*, Academic Press, Inc., New York, 1970 (ed. by G. E. Perlmann and L. Lorand), Vol. XIX, Article 70. Editors-in-chief S. P. Colowick and N. O. Kaplan.

15. Kelso, F. S., Matthews, J. M., and Kramer, H. P., *Anal. Chem.*, **36**, 577 (1964).

16. Szabo, J., and Joensuu, O., *Environ. Sci. Tech.*, **1**, 499 (1967).

17. Riley, J. P., and Taylor, D., *Anal. Chim. Acta*, **40**, 479 (1968).

18. Siegel, A., and Degens, E. T., *Science*, **151**, 1098 (1966).

19. Webb, K. L., and Wood, L., in *Automation in Analytical Chemistry, Technicon Symposium*, Mediad, Inc., New York, 1966, p. 440.

20. Dingman, J., Siggia, S., Barton, C., and Hiscock, K. B., *Anal. Chem;* **44**, 1351 (1972).

21. Dollman, G. W., *Environ. Sci. Tech.*, **2**, 1027 (1968).

22. Porter, C., Cahill, D., Schneider, R., Robbins, P., Perry, W., and Kahn, B., *Anal. Chem.*, **33**, 1306 (1961); see also Porter, C., and Kahn, B., *Anal. Chem.*, **36**, 677 (1964).

23. Strong, A. B., Rehnberg, G. L., and Moss, U. R., *Talanta*, **15**, 73 (1968).

24. Porter, C., Kahn, B., Carter, M. W., Rehnberg, G. L., and Pepper, E. W., *Environ. Sci. Tech.* **1**, 745 (1967).

25. Talvitie, N. A., *Anal. Chem.*, **43**, 1827 (1971).

26. Cranston, H. A., and Thompson, J. B., *Ind. Eng. Chem., Anal. Ed.*, **18**, 323 (1946).

27. Smith, J. B., in *Ion Exchangers in Organic and Biochemistry*, Interscience Publishers, Inc., New York, 1957 (ed. by C. Calmon and T. R. E. Kressman), Chap. 34.

28. Austerweil, G. V., in *Ion Exchangers in Organic and Biochemistry*, Interscience Publishers, Inc., New York, 1957 (ed. by C. Calmon and T. R. E. Kressman), Chap. 33.

29. Thompson, J., and Miller, V. J., *Plating*, **58**, 809 (1971).

30. Biechler, D. G., *Anal. Chem.*, **37**, 1054 (1965).

31. Freudiger, T. W., and Kenner, C. T., *Applied Spectroscopy*, **26**, 302 (1972).

32. Pitts, A. E., and Beamish, F. E., *Anal. Chem.*, **41**, 1107 (1969).

33. Sellers, N. G., *Anal. Chem.*, **44**, 410 (1972).

34. Volkin, E., Khym, J. X., and Cohn, W. E., *J. Am. Chem. Soc.*, **73**, 1533 (1951).

35. Khym, J. X., Doherty, D. G., and Cohn, W. E., *J. Am. Chem. Soc.*, **76**, 5523 (1954).

36. Dreze, A., Moore, S., and Bigwood, E. J., *Anal. Chim. Acta*, **11**, 554 (1954).

Ion-Exchange Chromatography

SEVEN

An attempt has been made to include in this chapter descriptions of chromatographic separations that occur principally through the single mechanism of ion exchange. These separations are governed chiefly by a diffusion process wherein there is a rapid stoichiometric exchange of one counter-ion for another. The affinity of each such counter-ion migrating down a column depends upon the magnitude of its charge and is also directly related to its mobility, which, in turn, is related to its size and shape; the ion exchanger shows a preference for the counter-ion with the smaller solvated volume. It is also tacitly assumed that in most cases the interstices or "pores" of the exchanger matrix do not mechanically exclude large ions by sieve action and that the main type of bonding between a counter-ion and the function group of the exchanger is ionic and not a physical sorbent force.

Deviations from the mechanism of "true" ion exchange are encountered less frequently in inorganic ion-exchange work in contrast to that found in work with organic substances. Such deviations are ignored here to avoid subdivision upon subdivision of subject matter, particularly where ion exchange appears to be the overwhelming mechanism involved in a separation. When ion exchange becomes less and less a dominant factor in separation work, or does not apply at all, but ion exchangers are still utilized as the sorbent materials, the work is given special attention in separate chapters (see Chapters 8 and 9). Ion-exchange chromatography of organic substances is considered in Part I, and inorganic substances in Part II of this chapter.

Part I
Ion-Exchange Chromatography of Organic Substances

A. AMINO ACIDS, PEPTIDES, AND PROTEINS

1. Amino Acids

Moore and Stein (1) demonstrated in 1950 that milligram amounts of amino acids could be quantitatively analyzed on columns (0.9 × 100 cm) of Dowex 50 (Na-form) by using 0.2 M sodium citrate buffers of progressively increasing pH from 3.4 to 11.0. The collected eluate fractions were analyzed manually with ninhydrin reagent for amino acid content. Since this demonstration, the efforts of a number of groups have transformed the original technique into one of the most reliable automated chromatographic systems that thus far exist. The original system of Moore and Stein was developed into a two-column automated procedure that is described by Spackman *et al.* (2a); this system requires dividing the sample on the two columns in order to complete an assay. The method has been modified by Spackman (2b) so that a mixture of amino acids, such as found in a protein hydrolyzate, may be analyzed in less than six hours (see Fig. 7-1). As illustrated on the chromatogram of Fig. 7-1, one column (0.6 × 10 cm) of this system is used to separate the basic amino acids and a longer column (0.9 × 58 cm) separates the acidic and neutral amino acids. A single-column automated amino acid analyzer has been developed by Hamilton (3) and has been refined by him and his co-workers so that in addition to analyzing complicated amino acid mixtures it is also possible to separate many of the constituents present in urine and other biological fluids. By this methodology more than 170 ninhydrin-positive substances in urine have been detected (4).

Both of the aforementioned chromatographic analyzers employed step-wise elution with buffer eluent solutions. Piez and Morris (5) designed a gradient-forming device that provides buffered eluent of increasing pH and ionic strength to a column. In single-column analyzers, this innovation gives better resolution of some peaks and sharpens the peaks towards the end of the chromatogram (where the elution of the basic amino acids occurs) so that quantitation of a peak is made easier.

All of these chromatographic systems have been improved in terms of their sensitivity, reliability, and completeness of automation.* These analyzers have been called upon to analyze a wide variety of samples. They have played an important role in the determination of the linear structure of proteins

---

*Different types and models of automatic amino acid analyzers are available commercially.

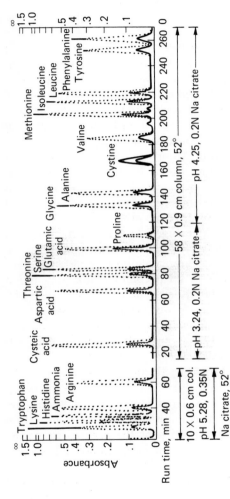

Fig. 7-1. Analysis of 100-nanomole synthetic mixture of amino acids. Conditions: columns filled with Beckman type UR-30 spherical cation exchanger, dimensions as shown; elution with sodium citrate buffers as shown at 50 ml/hr. (Reproduced from D. H. Spackman, *Methods of Enzymology*, Academic Press, Inc., New York, 1967 (ed. by C. H. W. Hirs), Vol. XI, p. 3, by permission of the author and Academic Press, Inc.)

Fig. 7-2. Amino acids found by ion-exchange chromatography in a single thumb print made on a wet glass surface. For visual comparison of peaks, the amount of lysine (0.011 μmole) is indicated on the figure. On the abscissa, 1 min is equivalent to 0.5 ml of column effluent volume. Conditions: column, 0.64 × 125 cm of Dowex 50-X8, 17.5 μ beads; elution with sodium citrate buffers (see reference 2) at 30 ml/hr. (Reproduced from P. B. Hamilton, *Nature*, **205**, 284 (1965), by permission of the authors and Macmillan Journals, Ltd.)

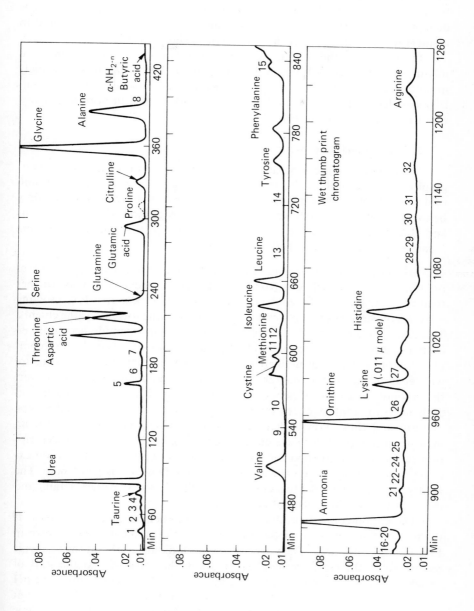

and polypeptides and are used routinely to screen physiological fluids and tissues for amino acid content. The sensitivity of these autoanalyzers is demonstrated by Hamilton's (6) determination of amino acids as found in a single wet thumb print. Such an analysis is shown in Fig. 7-2.

In addition to the routine examination of various complex amino acid mixtures, the automated amino acid analyzer is becoming an increasingly important tool for the accurate measurement of amino acids in very specialized samples related to the study of protein metabolism in normal and diseased persons (7, 8, 9, 10, 11).

2. Peptides and Proteins

Conditions for the separation of peptides by ion-exchange chromatography are not as well established as those that are routinely used for amino acid work. Each peptide mixture may present its own difficulty that can only be overcome by certain innovations of existing procedures. Only the resinous and the modified cellulose ion-exchange material are considered in the separation work described here.

(*a*) *Separations on resinous exchangers*   One of the main purposes of isolating individual peptides from mixtures is to recover them for further treatment or study. Schroeder (12) gives general procedures for resolving mixtures of peptides on polystyrene resins with volatile solvents as the eluents. Thus, the portion of the column effluent that contains peptide need only be evaporated to recover the material. Peptide mixtures are usually very complex solutions and it is unlikely that a single chromatographic procedure will separate the peptide mixture into all of its component parts. However, the amphoteric nature of peptides allows the use of both cation and anion exchangers to supplement any fractionation obtained on one or the other type of resin and thus allows different conditions of chromatography with an increase in capacity to achieve difficult separations (12). Only a brief description of the general method for the separation of peptides on the cation exchanger Dowex 50 and on the anion exchanger Dowex 1 is given here. The reader is directed to Schroeder's articles (12, 13) for the finer details and modifications of these general procedures.

The cation-column procedure employs gradient elution techniques. If the total volume of eluent is two liters or less, the gradient vessels shown in Fig. 7-3 are convenient. The mixer vessel has a diameter D and the reservoir section $1.4D$, the cross-sectional areas are in the ratio 1:2. At the start of an analysis, a given volume of pH 3.1 buffer (0.2 M pyridine adjusted to pH 3.1 with glacial acetic acid) is placed in the mixer chamber and twice that volume of pH 5.0 buffer (2 M pyridine adjusted to pH 5.0 with glacial acetic acid) is added to the reservoir cylinder. The sample is dissolved in water appropriate to the column size (see Table 7-1) and the pH is adjusted to 2

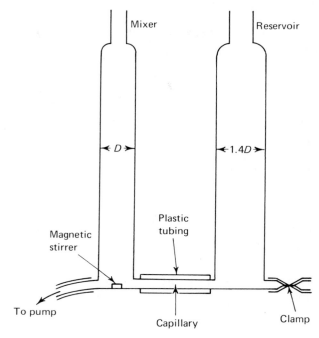

Fig. 7-3. Gradient vessels. (Reproduced from W. A. Schroeder, *Methods in Enzymology*, Academic Press, Inc., New York, 1972, ed. by C. H. W. Hirs and S. N. Timasheff, Vol. XXV, Part B, Article 15, by permission of the author and Academic Press, Inc.)

TABLE 7-1

Convenient Parameters for Various Sizes of Columns of Dowex 50-X2

Column size (cm)	Approximate ratio of column volumes	Equilibrating volume (ml)	Sample volume (ml)	Developer volume (ml) pH 3.1 buffer in mixer	Developer volume (ml) pH 5.0 buffer in reservoir	Fraction size (ml)	Rate of developer flow (ml/hr)
0.6 × 60		50	1	83	166	1	10
	1:4						
1 × 100		100	2	333	666	1.5	20
	1:2						
1.4 × 100		250	5	666	1332	3	30
	1:2.5						
2.2 × 100		500	15	1500	3000	5	60
	1:2.5						
3.5 × 100		1000	30	4300	8600	10	120

Reproduced from W. A. Schroeder, in *Methods in Enzymology*, Academic Press, Inc., New York, 1972 (ed. by C. H. W. Hirs and S. N. Timasheff), Vol. XXV, Part B, Article 15, by permission of the author and Academic Press, Inc.

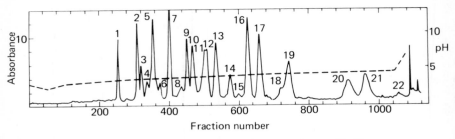

Fig. 7-4. Separation of peptides in a tryptic hydrolyzate on Dowex 50. (Reproduced from W. A. Schroeder, R. T. Jones, J. Cormick, and K. McCalla, *Anal. Chem.*, **34**, 1570 (1962), by permission of the authors and the American Chemical Society.)

with hydrochloric acid and then the peptide mixture is carefully layered on the column before the gradient device is started. The color reaction with ninhydrin may be applied to determine in which fractions peptides have emerged. This may be done manually or the photometric determination may be automated (13). By these techniques, a separation such as that shown in Fig. 7-4 can be achieved.

Once simpler mixtures are obtained by the cation-exchange procedure, further resolution of the mixtures may be carried out, if necessary, on the anion exchange resin Dowex 1. The separation of such simpler mixtures may be achieved on smaller columns at considerably higher sample loadings. Eluent is supplied to these columns after a series of solvents are introduced successively to a constant volume mixer according to the schedule shown in Table 7-2. The solvent changes may be made manually or by automated devices. A peptide fraction obtained from a cation column according to the

TABLE 7-2

Volume of Developers for Dowex 1 Chromatograms

Developer	Column 0.6 × 60 cm (ml)	Column 1 × 100 cm (ml)
Buffer pH 9.4	10	40
Buffer pH 8.4	30	120
Buffer pH 6.5	40	160
Acetic acid 0.5 *N*	60	240
Acetic acid 2 *N*	100	400

Reproduced from W. A. Schroeder, in *Methods in Enzymology*, Academic Press, Inc., New York, 1972 (ed. by C. H. W. Hirs and S. N. Timasheff), Vol. XXV, Part B, Article 16, by permission of the author and Academic Press, Inc.

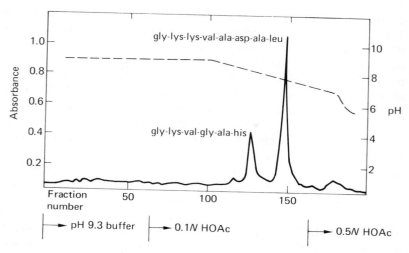

Fig. 7-5. Separation of a peptide fraction on Dowex 1 beginning with pH 9.3 buffers. Peptide fraction obtained from about fraction 540 from the Dowex 50 column of Fig. 7-4. (Reproduced from W. A. Schroeder, R. T. Jones, J. Cormick and K. McCalla, *Anal. Chem.*, **34**, 1570 (1962), by permission of the authors and the American Chemical Society.)

conditions of Fig. 7-4 is further resolved by rechromatography on the anion exchanger Dowex 1 as is shown in Fig. 7-5.

Large peptides or highly basic proteins that are not readily fractionated on the Dowex type resins may, in some instances, be separated on Amberlite IRC-50, a weakly acidic cationic copolymer of methacrylic acid and divinylbenzene. This resin has a pK_a of about 6.0, therefore the carboxylic groups are almost fully ionized above 7 and nonionized below 5.0 (14). For columns operated below pH 6.5, nonionic interactions of the peptide or protein and of the acid form of the resin may contribute to chromatographic separation and at pH's above 6.5 both ion exchange and physical sorption forces may be involved in the resolution of peptide mixtures. Conditions for the separation of a variety of peptides are given by Edmundson (14) and by Moore and Stein (15). The subject of separating peptide mixtures on carboxylic resins is not treated further, since such mixtures are now commonly resolved on cellulosic ion-exchange materials.

(*b*) *Separations of peptides and proteins on cellulosic ion exchangers*
The separation of polyelectrolytes such as high molecular-weight peptides and proteins on ion-exchange materials is a unique type of chromatography in that a large number of bonds need to be formed and broken during the chromatographic process. Due to irreversible or very tight bonding of high molecular-weight peptide oligomers and proteins to resinous exchangers,

chromatography on these materials is very unsatisfactory (see Chapter 3). These limitations to chromatography have been overcome by the incorporation of ionic groups into the cellulose matrix, thus creating ion-exchange materials that have become very powerful tools for the fractionation of polyelectrolytes. Peterson and Sober (16) first introduced, as an ion-exchange material, chemically modified cellulose that had diethylaminoethyl groups linked to the hydroxyl groups of the cellulose. Since then a variety of cellulosic ion exchangers have been synthesized. Descriptions of these materials as well as techniques for preparing them for column use can be found in reports by Roy and Konigsberg (17), Peterson (18), Fasold *et al.* (19); also see Chapter 3 of this book. Only a few examples of the fractionation of peptides or proteins on cellulosic ion exchangers will be cited here.

Carboxymethyl cellulose (CM-cellulose), a weakly acidic cation exchanger, has been used by Clegg *et al.* (20) to separate the α and β polypeptide chains of human hemoglobin. Essentially this same technique was employed by Popp and Bailiff (21) to separate the α and β chains of mouse hemoglobin prior to sequence determinations of amino acids in the major and minor β

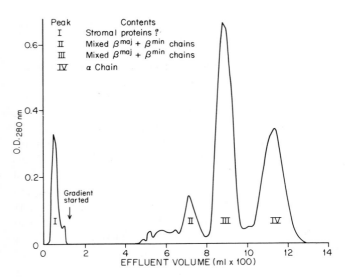

Fig. 7-6. Separation on Whatman CM23 of α- and β-chain polypeptides of BALB/c mouse hemoglobin. Column size was 1.9 $\times$ 33 cm. Starting buffer contained 8.0 M urea, 0.05 M β-mercaptoethanol, and 5 mM Na_2HPO_4 adjusted to pH 6.8 with H_3PO_4. The arrow indicates where the arithmetic gradient with 0.03 M Na_2HPO_4 buffer (pH 6.8) was started; the reservoir and mixing chamber each contained 600 ml. The optical density of the effluent at 280 nm is plotted against the effluent volume. (Reproduced from R. A. Popp and E. G. Bailiff, *Biochim. Biophys. Acta.*, **303**, 61 (1973), by permission of the authors and Elsevier Publishing Company.)

chains of the hemoglobin. The separation of polypeptides of mouse hemoglobin is illustrated in Fig. 7-6. Various operational parameters such as changes in flow rates, salt concentrations, gradient volumes, and sample loadings have been studied by Smith and Stahman (22) for the purpose of finding optimum conditions for the separation of lysine polypeptides on CM-cellulose.

In studies elucidating the active site areas of the enzyme α-chymotrypsin, Nakagawa and Bender (23) methylated the histidine-57 residue of this enzyme with radioactive methyl p-nitrobenzenesulfonate. Subsequent cleavage of the disulfide bond of [^{14}C] methylchymotrypsin produced the B and C chains of this enzyme as separate polypeptides. Separation of these chains by chromatography on diethylaminoethyl cellulose (DEAE-cellulose) (see Fig. 7-7), a weakly basic anion exchanger, revealed that the main radioactive peak corresponded to the main protein peak of the B chain as measured by ultraviolet absorption. The conclusion of this experiment is that the radioactive methyl group is in the B chain. This was confirmed by analysis of the fragments produced by enzymatic hydrolysis, which indicated that mainly the histidine-57 residue was modified.

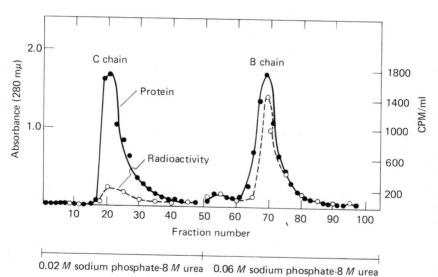

Fig. 7-7. Separation of B and C chains of [^{14}C] methyl-modified chymotrypsin. Chains B and C were separated on a DEAE-cellulose column (3.5 × 25 cm) by stepwise elution with 0.02 M sodium phosphate (pH 7.91)-8 M urea, and 0.06 M sodium phosphate (pH 7.89)-8 M urea. The 10-ml fractions were collected in tubes and the flow rate was 60 ml/hr. An aliquot of 250 μl of eluent of each test tube was used for the radioactivity measurement. (Reproduced from Y. Nakagawa and M. L. Bender, *Biochemistry*, **9**, 259 (1970) by permission of the authors and the American Chemical Society.)

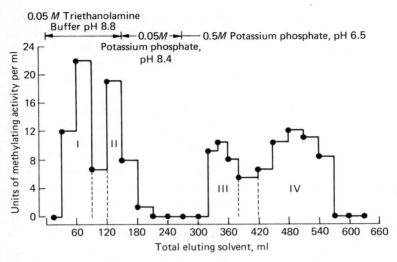

Fig. 7-8. Chromatography of methylating enzymes on phosphocellulose. Chromatography was carried out as described in the text. (Reproduced from J. Hurwitz, M. Gold, and M. Anders, *J. Biol. Chem.*, **239**, 3462 (1964), by permission of the authors and the American Society of Biological Chemists, Inc.)

Cellulose phosphate (P-cellulose), a strong acid exchanger, was used by Hurwitz *et al.* (24) to purify and isolate enzyme fractions that catalyze the methylation of soluble ribonucleic acid. A dialyzed ammonium sulfate fraction was diluted with 0.05 *M* triethanolamine buffer, pH 8.8, and poured onto a column of P-cellulose (3 × 30 cm) that had been previously washed with the same 0.05 *M* buffer. Stepwise elution of protein was made with the buffers as indicated in Fig. 7-8.

B. SEPARATION OF CARBOHYDRATE SUBSTANCES

1. Neutral Carbohydrates

Most neutral carbohydrates form anionic complexes with borate. This reaction is the basis of a method for the chromatography of sugars on anion-exchange resins. For the separation of a group of similar carbohydrate-borate complexes, presumably representative of the type (25, 26) it was necessary to find sets of conditions for which the different borate diols exhibited a different degree of affinity for the exchanger. Such conditions were found for a variety of saccharides by Khym and Zill (26–28) and their co-workers utilizing anion exchangers in their borate form and dilute solutions of sodium or potassium tetraborate of pH's between 7 and 9 as eluents. From the chromatographic behavior of a number of different types of carbohydrate compounds, the following general rules became apparent (29).

1. *Cis*-1,2-glycols are strongly attracted to the borate exchanger; *trans*-1,2-glycols are not.

2. *Cis*-1,2-glycols in a pyranose system show little affinity toward the borate exchanger relative to those in a furanose system, which show a high affinity for the exchanger.

3. The greater the number of *cis*-1,2-glycols in a compound, the higher is its affinity.

4. Reducing sugars in which the hydroxyl group on C_1 is unsubstituted can form additional *cis*-1,2-glycols by mutarotation and hence show an increased attraction for the borate exchanger.

The chromatographic separations just discussed were carried out on simple mixtures prepared by adding, to dilute borate solutions, milligram amounts of from three to five reference compounds. The presence of carbohydrate in effluents was determined colorimetrically (manually) on fractions eluted from the exchanger. Such separations required large volumes of eluent even though small columns were employed (26–28). This is a disadvantage in two respects: (1) the system is impractical for the analysis of small (microgram) amounts of carbohydrate; and (2) the length of analysis, up to 60 hours in some cases, is prohibitory for routine work.

Various elaborations (30–35) of the borate process for separating polyhydroxy compounds have overcome many of the disadvantages of the original investigations. To minimize alkaline degradations and rearrangements of sugars during chromatography, particularly at elevated temperatures, elutions are now commonly performed with borate buffers of pH 7. To increase resolution and sensitivity, and to shorten the time of analysis, columns generally are now packed with finer resin particles, operated at temperatures above ambient, and the eluents are applied to a column with gradient elution devices. The effect of employing some of these newer operating parameters can be seen by the separation shown in Fig. 7-9, wherein the narrow elution peaks were obtained with very small volumes of eluent in comparison to those obtained in the earlier separations mentioned (26–28).

The trend in this type of methodology for the analysis of carbohydrate mixtures is towards complete automation of the chromatographic procedure. Green (36) and Scott *et al.* (37) designed prototype systems for automated, high resolution analyses of carbohydrate materials. Improvements and innovations of the basic system have been made by Kesler (32), Walberg

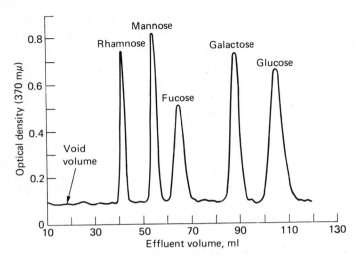

Fig. 7-9. Column chromatographic separations of a standard neutral mono-saccharide mixture containing 0.40 μmole rhamnose, 0.75 μmole mannose, 0.50 μmole fucose, 1.00 μmole galactose, and 1.25 μmoles glucose. Conditions: column, 0.6 × 155 cm Dowex 2-X8, 200–400 mesh, chloride-borate form; elution at 50°C with 0.4 M boric acid, 1.0 M glycerol, 0.050 M NaCl, adjusted to pH 6.8; flow rate, 3.0 ml/hr. Fractions were analyzed for sugar using the aniline/acetic orthophosphoric acid method. (Reproduced from E. F. Walborg, Jr., L. Christensson, and S. Gardell, *Anal. Biochem.*, **13**, 177 (1965), by permission of the authors and Academic Press, Inc.)

et al. (31a-c), Ohms *et al.* (34), and Lee *et al.* (33). These automatic systems* are all very similar to the design that is schematically illustrated in Fig. 7-10. Floridi (34a) ignored the trend towards gradient elution systems and showed that in his automated system a stepwise elution technique gives complete separation of several mono-, di-, and trisaccharides present in complex mixtures of sugar samples. By using only borate in the eluting buffers, Floridi (34a) eliminated long reequilibration periods following a determination. Hough *et al.* (35) utilized commercial equipment to analyze carbohydrate mixtures derived from glycoproteins and polysaccharides. These workers made several additional improvements in operative techniques. These are: (1) the detection of carbohydrates by the cysteine-sulfuric acid assay, which has many advantages over other methods (35); (2) gradient elution of the column with borate buffers of pH 7 produced by a multichambered device, which gave good resolution without the inclusion of glycerol (31b) or butane-2,3-diol (31a, 31c), substances that interfere with convenient colorimetric

*A commercial carbohydrate autoanalyzer is marketed by the Technicon Instrument Company of New York.

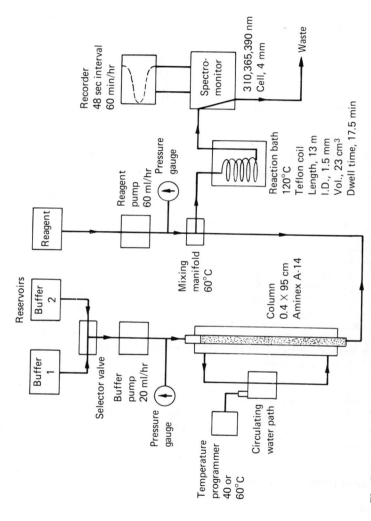

Fig. 7-10. Schematic diagram of an automated carbohydrate analyzer. (Reproduced from E. F. Walborg, Jr., and L. E. Kondo, *Anal. Biochem.*, **37**, 320 (1970), by permission of the authors and Academic Press, Inc.)

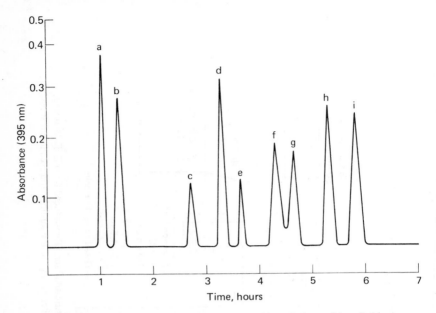

Fig. 7-11. Separation of various carbohydrates: (a) trehalose, (b) cellobiose, (c) L-rhamnose, (d) D-ribose, (e) D-mannose, (f) L-arabinose, (g) D-galactose, (h) D-xylose, (i) D-glucose; 560 nmoles of each carbohydrate were used with the exception of trehalose (284 nmoles) and cellobiose (355 nmoles). Conditions: column, 0.6 × 75 cm of type-S, Technicon Chromobeads, chloride-borate form; elution with a borate-chloride gradient at 53.5°C; flow rate 37 ml/hr; the column eluate was analyzed for carbohydrates by reaction with a solution of cysteine hydrochloride [0.7% w/v in sulfuric acid (86% v/v)]. L. Hough, J. V. S. Jones and P. Wusteman, *Carbohyd. Res.*, **21**, 9 (1972), by permission of the authors and Elsevier Publishing Company.)

reactions; (3) the essential elimination of alkaline degradations of sugars during chromatography (31c, 38) even though the gradient employed is operated at elevated temperatures. The separation shown in Fig. 7-11 is typical of those obtained by the automated procedures as used by Hough *et al.* (35).

2. Carbohydrate Derivatives

Sodium tetraborate ($Na_2B_4O_7$) or potassium tetraborate ($K_2B_4O_7$) at low concentrations (10^{-5} to 10^{-1} M) in eluting solutions of pH 8–10 are effective in chromatography involving the borate diols of substances that have additional negative charges, such as the nucleotides (39) and the sugar phosphates (40). In these instances, the exchangers involved are converted to salt forms other than borate. However, borate is present in the eluent but only at concentrations sufficient to allow borate complexing. The other ions present in the eluents, such as chloride or sulfate, effect the migration of the borate diols down the chromatographic column. Weaker anionic substances, such as the ribonucleosides, are chromatographed on the borate form of exchangers (25, 41), as are the sugars.

Aldobiouronic acids were separated in borate medium by Samuelson and Wictorin (42). Dowex 1-X8 (0.3 × 87 cm) was used in the borate form and elutions were carried out with 0.12 M tetraborate solution. Borate complexing was also used by Schenker and Rieman (43) to determine malic, tartaric, and citric acids in fruit samples by anion-exchange chromatography. Samples were separated on 3.8 sq. cm × 25 cm of Dowex 1-nitrate with sodium nitrate-borate solutions of pH 6.15 as the eluents.

C. ORGANIC ACIDS

1. Introduction

The first practical demonstration that a complex mixture of organic acids could be resolved on an anion exchanger came from the work of Busch *et al.* (44) who eluted acids of the citric acid cycle (and related acids) with increasing concentrations of formic acid from a Dowex 1 anion column in the formate form. This work was also important in that it introduced the concept of gradient elution to ion-exchange chromatography. Since then, anion-exchange chromatography has been applied to the separation of the organic acids not only belonging to the Krebs cycle but in general to the members belonging to a homologous series such as the acids classified as mono-, di-, or tricarboxylic, aldehydic or ketonic, uronic or aldonic, hydroxy, and unsaturated. Some of this work will be examined here.

2. Detection Methods

Until recently, lack of sensitive and convenient methods for detecting organic acids in column eluates has been a deterrent to progress in this field of ion-exchange technology. A brief description of the developments in this area is given here instead of elaborating on the analytical detection schemes that relate to the subject matter in the sections that follow.

In the earlier publications, cumbersome manometric, titrimetric (44), or colorimetric techniques (44, 45,) were applied to column eluates. Many of these are too specific and are not applicable to the detection of the many different classes of organic acids, or are not amenable to the analysis of a large number of samples. Samuelson and his collaborators (46–48, 50–54) developed equipment and investigated chemical reactions that are capable of determining organic acids by automated methodology. Chromic acid oxidation is applied as a nonspecific method that is based upon the determination of green chromium(III) ions subsequent to the oxidation of an organic acid. The carbazole reaction is applied to the detection of the uronic acids and some other hydroxy acids. A third method detects formaldehyde that arises from the oxidation with periodate of an organic acid that contains a primary alcohol group. A fourth procedure measures the consumption of periodate by an organic acid through ultraviolet spectrophotometry. The Swedish group has developed instrumentation whereby column eluates are subjected to analysis in a four-channel analyzer sensitive to the methods just discussed (48).

A more recent advance in organic acid detection comes from the work of the Body Fluids Analysis Program of the Oak Ridge National Laboratory. A miniature flow fluorometer developed by Thacker (55) has been used by Katz *et al.* (56) to monitor column effluents for organic acids. In this application, the instrument detects fluorescent cerium(III) produced when compounds such as organic acids react with cerium(IV) as this reagent is injected automatically into eluate streams.

Ascorbic, dehydroascorbic, and diketogulonic acids, separated on an anion exchanger by Hegenauer and Saltman (57) were detected in column eluates by manual procedures. The three acids were determined by absorbance measurements in the ultraviolet as well as by the colorimetric measurement of their 2,4-dinitrophenylosazones.

3. Uronic and Aldonic Acids

Methods for the analysis of uronic acids in biological material have always been rather limited due to the interference caused by sugars. Khym and Doherty (45) demonstrated that galacturonic and glucuronic acids may be separated on the acetate form of Dowex 1 with 0.15 *M* acetic acid as the eluent. Neutral sugars such as arabinose and galactose pass through the acetate exchanger in the absence of borate while the two uronic acids are retained and may be subsequently separated as discrete solute bands. Presently, very complex mixtures of uronic acids are efficiently analyzed. This is due to the efforts of Samuelson and his co-workers who have done extensive research in developing methods for the ion-exchange chromatography of many different classes of organic acids. Their studies in this area have led to a method for the automated analysis of complex mixtures of uronic acids by anion-exchange chromatography with sodium acetate and acetic acid solutions as eluents (46, 47, 48). The separation of a complex mixture of uronic acids by the automated procedure of Johnson and Samuelson (46) is shown in Fig. 7-12. This work was extended by this group (47, 48) to include a large number of other acids such as the aldonic, saccharinic, and aliphatic keto acids. In this latter effort, the chromatographic separation of 44 organic acids, mainly hydroxy acids, has been studied using the acetate exchange system of Fig. 7-12.

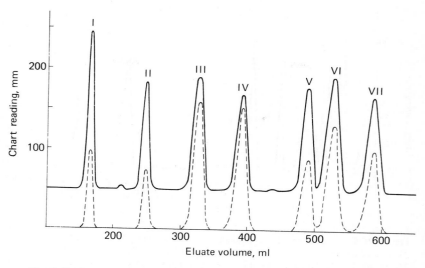

Fig. 7-12. Separation of 6-O-(β-D-glucopyranosyluronic acid)-D-galactose (I), cellobiouronic acid (II), 4-O-methyl-D-glucuronic acid (III), D-galacturonic acid (IV), L-guluronic acid (V), D-glucuronic acid (VI) and D-mannuronic acid (VII). Carbazole method: dotted line; dichromate method: full line. Eluant: 0.05 *M* sodium acetate, pH 5.9; flow rate: 1.06 ml/min. Column, 0.6 × 135 cm, Dowex 1-formate, 26–32 μ, operated at 30°C. (Reproduced from S. Johnson and O. Samuelson, *Anal. Chim. Acta*, **36**, 1 (1966), by permission of the authors and Elsevier Publishing Company.)

Fransson *et al.* (49) modified a commercially obtained sugar autoanalyzer so that it is suitable for the automated determination of uronic acids and uronic acid-containing oligosaccharides. Samples containing 50–200 μg of uronic acid, or its equivalent as a derivative, were applied in water solution (1–2 ml) on a column (0.63 × 140 cm) of Dowex 1-X8, 200–400 mesh, formate form. The column was eluted with 0.5 *M* formic acid at a rate of 30 ml per hr, which required a pressure of 200–300 psi. The effluent from the column was analyzed by the automated orcinol-H_2SO_4 method.

4. Other Aliphatic Carboxylic Acids

In addition to investigations of monocarboxylic acids, Samuelson and his collaborators have extensively studied ion-exchange chromatographic schemes for the quantitative determination of di-and tricarboxylic acids and of straight- and branched-chain monocarboxylic acids. The elution behavior of the former two classes of organic acids was studied in magnesium acetate (50), sodium sulfate (51, 52), and sodium phosphate solutions (52, 53). The branched and unbranched monocarboxylic acids were eluted with sodium acetate-acetic acid buffers (54). These studies were carried out on

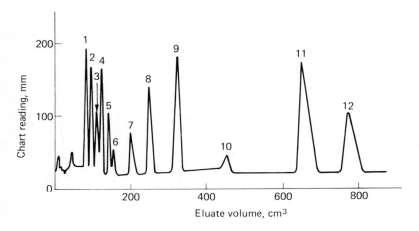

Fig. 7-13. Separation of 50 μg oxalacetic (pyruvic) (1), 25 μg galactaric (2), 25 μg glucaric (3), 25 μg malic (4), 25 μg L(+)-tartaric (5), 200 μg malonic (6), 250 μg oxalic (7), 50 μg itaconic (8), 50 μg α-ketoglutaric (9), 100 μg isocitric (10), 100 μg maleic (11) and 50 μg citraconic acid (12). Eluent, 0.4 M sodium phosphate; pH, 5.5; flow rate, 6.4 cm min⁻¹; resin bed, 4.3 × 1410 mm. (Reproduced from L. Bengtsson and O. Samuelson, *Anal. Chim. Acta*, **57**, 93 (1971) by permission of the authors and Elsevier Publishing Company.)

analytical columns coupled with automatic analysis of the eluates. A complex mixture of dicarboxylic acids as separated in phosphate media is shown in Fig. 7-13.

To investigate the chelating properties of ascorbate and its oxidation products and hydrolytic derivatives, Hegenauer and Saltman (57) developed a rapid quantitative system for unambiguously separating mixtures of ascorbic, dehydroascorbic, and diketogulonic acids. Samples containing 25–50 μmoles of each component were analyzed on a column (0.9 × 56 cm) of Dowex 1-X2 using continuous elution with a single eluent, 0.05 M H_3PO_4 at 30 ml per hr (room temperature).

5. Separation of Aromatic Acids

Katz *et al.* (56) developed new monitoring techniques that detect organic acids in column effluent streams. Krebs-cycle acids as well as cerate-oxidizable aromatic acids are now detected among the perhaps hundreds of molecular constituents found in complex biochemical mixtures (body fluid samples) that were previously analyzed only for UV-absorbing constituents. The synthetic mixture shown separated in the chromatogram of Fig. 7-14 demonstrates that many aromatic organic acids, previously undetected by less sensitive UV-monitoring devices, appear as discrete peaks when the eluate is also assayed with cerate reagent.

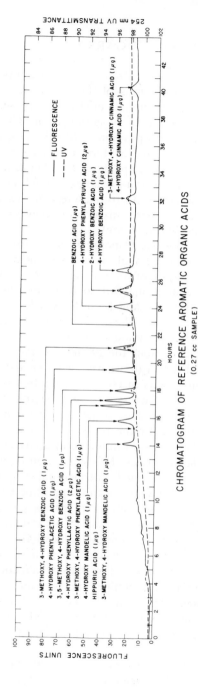

Fig. 7-14. Separation of aromatic acids by anion exchange. Conditions: column, 0.22 × 150 cm, Bio-Rad Aminex A-27, 12.5 to 13.5 μ beads. Elution at 5.5 ml/hr with an ammonium acetate-acetic acid buffer solution of pH 4.4 having an acetate concentration that gradually increases from 0.015 M to 6.0 M. Column operated at ambient temperatures for 10 hours and at 60°C thereafter. Under these conditions pressure required, 1500–2000 psi. Parallel monitoring of column eluate by UV and cerate oxidative responses. (Reproduced from S. Katz, W. W. Pitt, Jr., and G. Jones, Jr., *Clin. Chem.*, **19**, 817 (1973), by permission of the authors and the American Association of Clinical Chemists.)

D. AMINO SUGARS

Gardell (58) was the first to demonstrate the applicability of cation-exchange chromatography for the analysis of amino sugars such as glucosamine and galactosamine. These simple mixtures were separated on Dowex 50 with 0.3 N hydrochloric acid as the eluting agent. Crumpton (59) extended this work and lists the elution position of 10 amino sugars as separated on Zeo-Karb 225 with 0.33 N hydrochloric acid as the eluent. Galactosamine, gulosamine, and allosamine are eluted together as a single peak in this system. From this point on, the ion-exchange technology in this field grew rapidly as investigators sought ways to analyze complex mixtures of amino sugars and also to analyze these mixtures as they are present in hydrolyzates of protein-polysaccharide complexes.

Since amino sugars give a positive response to ninhydrin reagent, the equipment designed for the automated analysis of amino acids can be used in a similar manner for the analysis of amino sugars. Brendel *et al.* (60) modified the Technicon amino acid analyzer to carry out such analysis. Their paper describes separations obtained on Dowex-50 columns with pyridine-acetic acid buffers as eluents. These buffers were able to resolve complex mixtures of amino acids and amino sugars including some diamino sugars and amino hexuronic acids among the 29 amino compounds studied. Lee *et al.* (61) and Fanger and Smith (62) developed similar methodology for the determination of amino sugars in glycoproteins. In the former system mixtures of glucosamine and galactosamine may be analyzed in about 90 minutes. In this system as little as 1 μg of hexosamine is detected by a colorimetric method not influenced by the presence of other amino compounds. Yaguchi and Perry (63) used an amino acid analyzer to separate seven of the possible eight 2-hexosamines. This separation is carried out in the presence of borate at pH 7.4. Although not mentioned by the authors, borate complexing may have played a role in this separation since the positive charge of the amino group in these compounds may be partially neutralized, each to a different degree, by the formation of anionic borate diol complexes. This may be more apparent from the work of Gregory and Van Lenten (64) who separated the 2-amino-2-deoxy isomers in phosphate media and also in borate media alone. As seen in Fig. 7-15, which compares these two eluent systems, mannosamine, talosamine, and lyxosamine elute much sooner with borate as the eluent. This indicates the acquisition of negative charges due to borate diol formation in some of these 2-aldosamines, hence aiding in the early desorption of these amino compounds from the cation exchanger.

Lee (65) describes in detail in *Methods in Enzymology* (Volume XXVII) an autoanalytical system designed to determine amino sugars.

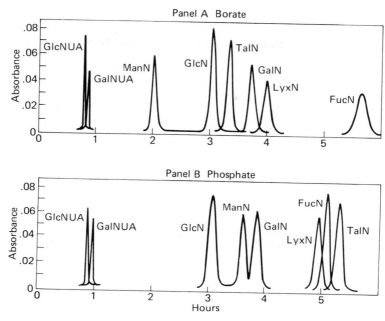

Fig. 7-15. Separation of 2-amino-2-deoxy-D-hexoses. Columns 0.6 × 130 cm, Type A chromobeads (Technicon resin), operated at 60°C. Elution: Panel A, phosphate buffer at 1 ml/min. Panel B, borate buffer at 1 ml/min. GlcNUA (glucosamineuronic acid), GalNUA (galactosamineuronic acid), ManN (mannosamine), GlcN (glucosamine), TalN (talosamine), GalN (galactosamine), LYXN (lyxosamine), FucN (fucosamine). (Reproduced from J. D. Gregory and L. V. Lenten, in *Automation in Analytical Chemistry*, *Technicon Symposium 1965*, Mediad, Inc., New York, 1966, p. 620, by permission of the authors and Mediad, Inc.)

E. AMINES

Various classes of amines are separable on weak-acid cation cellulosic exchangers as well as on the weak- and strong-acid resinous types. Tompsett (66) took advantage of the high affinity of polystyrene resins for aromatic substances to separate a group of primary aromatic amines from other materials present in biological samples. A strong-acid cation exchanger was used in this case. More complicated mixtures containing aliphatic, aromatic, and heterocyclic amines were separated on a similar exchanger by Hatano *et al.* (67). In this case, the separations were carried out with the aid of automated chromatographic equipment originally designed for amino acid work. According to this procedure, 0.1 to 1.0 μmole amounts of amines present in mixtures containing as many as 16 species are automatically analyzed to an

TABLE 7-3

Compositions and Conditions of Eluting Buffers for the Chromatogram of Fig. 7-16

Buffer system	Concn., N	Benzyl alcohol concn. (%)	pH[a]	Ionic strength of sodium ion[b] M
I. Sodium citrate	0.116	—	5.28	0.35
II. Sodium borate	0.025	0.5	8.02	0.60
III. Gradient				
1. Sodium borate	0.05	0.4	10.00	0.60
2. Sodium salicylate	0.20	—	11.50	0.65
3. Sodium salicylate	0.20	—	12.50	0.70

[a]Adjusted with 6 *N* sodium hydroxide or hydrochloric acid solution.
[b]Adjusted with sodium chloride solution.
Source: Reproduced by H. Hatano, K. Sumizu, S. Rokushika, and F. Murakami, *Anal. Biochem.*, **35**, 377 (1970), by permission of the authors and Academic Press, Inc.

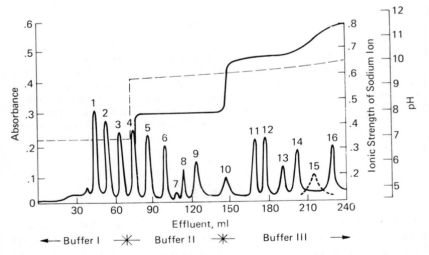

Fig. 7-16. Chromatogram of authentic mixture of sixteen amines and eluting conditions of the buffer system. Peaks: (1) methylamine, (2) ethylamine, (3) allylamine, (4) *n*-propylamine, (5) isobutylamine, (6) *n*-butylamine, (7) 1, 2-propanediamine, (8) histamine, (9) isoamylamine, (10) *n*-amylamine, (11) tyramine, (12) putrescine, (13) phenethylamine, (14) cadaverine, (15) serotonin, (16) hexamethylenediamine. The amount of every amine was 0.4 μmole except for serotonin, 1.0 μmole. Column, 0.6 × 12 cm, Aminex A-4 (Bio-Rad Laboratories) operated at 50°C. Elution as indicated at 30 ml/hr. (Reproduced from by H. Hatano, K. Sumizu, S. Rokushika, and F. Murakami, *Anal. Biochem.*, **35**, 377 (1970), by permission of the authors and Academic Press, Inc.)

accuracy of $\pm 3.5\%$ within eight hours. Three different eluents are supplied to the column. The first two are applied in stepwise fashion, and the third by a gradient device (a three-component gradient system). The composition of the eluting buffers is shown in Table 7-3 and the schedule for supplying them to the column is given in Fig. 7-16, which shows the chromatogram of a mixture of 16 amines. A similar automated method was developed by Morris *et al.* (68) for analysis of polyamines. By this procedure arginine, putrescine, spermidine, and spermine are widely separated on a polyacrylic cation exchanger (Bio-Rex 70) using pyridine-acetate solutions as eluents. The chromatographic separation and quantitation of biogenic amines that arise in physiological fluids has been made by Wall (69). In these studies Zeo-Karb 226, a carboxylic type exchanger, was packed into columns coupled to equipment capable of automatically monitoring column effluents. Citrate buffers containing propanol were utilized as eluents and, as seen in Fig. 7-17, this allowed the separation of mixtures containing biogenic amines.

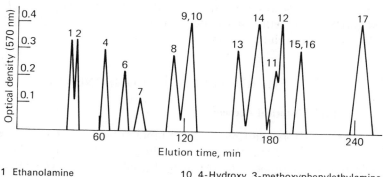

1 Ethanolamine	10 4-Hydroxy, 3-methoxyphenylethylamine
2 Ammonia	11 Putrescine
3 Epinephrine	12 Cadaverine
4 Norepinephrine	13 Serotonin
5 1-Aminobutane	14 Phenylethylamine
6 Octopamine	15 Agmatine
7 Dopamine (oxidized and cyclized)	16 Histamine
8 Tyramine	17 Tryptamine
9 β-Hydroxyphenylethylamine	

Fig. 7-17. Ion-exchange chromatography of some biogenic amines, column, 0.63×24 cm, Zeo-Karb 226, 21 μ beads, operated at 60°C. Gradient elution with sodium citrate-propanol solutions (see reference 69) at 30 ml/hr. Compounds 3 and 5 (see list above) were not included in this chromatogram. (Reproduced from R. A. Wall, *J. Chromatog.*, **60**, 195 (1971), by permission of the authors and Elsevier Publishing Company.)

Lepri *et al.* (70) have demonstrated that primary aromatic amines such as *o*-nitroaniline, *o*-aminobenzoic acid, *p*-nitroaniline, sulfanilamide, *p*-aminohippuric acid, and *p*-aminobenzoic acid are separated on carboxymethylcellulose columns having a cross section of 0.94 cm² and filled with about 2 g of the acid form exchanger. Water alone and acetic acid solutions served as the eluents.

F. SEPARATION OF CARBONYL COMPOUNDS IN THE PRESENCE OF BISULFITE

Carbonyl compounds form negatively charged species in the presence of bisulfate ion. Christofferson (71) took advantage of this reaction to determine acetaldehyde, formaldehyde, pyruvic acid, pyruvaldehyde, 5-hydroxymethylfurfural, furfural, and vanillin in sulfite-spent liquors. The compounds are taken up from the liquor on an anion exchanger in the bisulfite form and separated by elution with sodium bisulfite solutions containing ethanol. The seven compounds are separated in the order given above on a 1.0 × 8.0 cm column of Dowex 1-X8 at a flow rate of about 0.72 cm per min using 0.4 M NaHSO$_3$-20% ethanol to elute the first six compounds and 0.8 M NaHSO$_3$ -20% ethanol to elute vanillin.

G. NUCLEIC ACID COMPONENTS

1. Brief Historical Account

Three factors contributed to the rapid growth of ion-exchange technology in this field. First, there existed a huge backlog of problems in nucleic acid research waiting to be solved by the advent of newer methods for the separation and isolation of various nucleic acid components. Second, there existed a simple, sensitive, and nondestructive assay method for these components, namely direct estimation by ultraviolet spectrophotometry. Third, various chemically stable, high capacity, synthetic ion-exchange resins became commercially available. It had already been established that these exchangers could separate closely related species such as the rare earths (72). Shortly thereafter, the resinous exchangers were introduced into amino acid work (see Section A) and at about the same time, Cohn (73) clearly demonstrated the value of these polystyrene-divinylbenzene exchangers for the separation, isolation, and analytical determinations of nucleic acid fragments. Following Cohn's landmark experiments, applications of ion-exchange chromatography to nucleic acid work flourished. Several trends in this field then developed.

As the limitations of stepwise elution became apparent, gradient elution methods were introduced into column procedures (74). This new technique greatly improved resolution of the known compounds and revealed several new nucleic acid components. The practical advantage gained by gradually increasing the ionic strength of the eluent during chromatography is seen in the original experiments of Hurlbert et al. (75) who separated the acid-soluble nucleotides on anion exchangers by this technique. Another example of the effectiveness of gradient elution is shown by the narrow solute bands of purines and pyrimidines that were obtained from cation exchangers by Crampton et al. (76) who utilized increasing concentrations of sodium citrate or of ammonium formate of pH 4 to effect such separations.

The resin exchangers themselves became limited for use as ion-exchange materials, as researchers turn their attention to the separation and isolation of the larger molecular-weight nucleic acid fragments such as the polynucleotides. The problem of difficult elution or irreversible binding of some oligonucleotides to polystyrene resins was circumvented by the experiments of Staehelin et al. (77) and Tener et al. (78). These workers demonstrated that synthetic homologous oligonucleotides are easily separated on cellulose exchangers according to their chain length. The inability to separate the larger mixed oligonucleotides containing both purine and pyrimidine bases led to the important finding of Tomlinson and Tener (79) who showed that mixed polynucleotides could be separated on DEAE-cellulose when eluents contained 7 M urea. The presence of urea in the eluents suppressed secondary binding forces (H-bonding) affecting the column chromatography of the polynucleotides and thus allowed separations to take place. Staehelin (80) describes in detail these implications of polynucleotide behavior and gives a general review on the separation of oligonucleotides and polynucleotides on cellulosic exchangers.

As the technology improved and allowed for the isolations of homogeneous nucleotide polymers, it became apparent that if linear sequence determinations were to be made on these polymers, methods for detecting low molecular-weight constituents (e.g., mononucleotides, nucleosides, and purine and pyrimidine bases) at very low levels of concentrations would be needed. This fact and the intense interest in detecting metabolites at extremely low levels in physiological fluids gave the impulse for the reexamination of existing chromatographic procedures for the purpose of modifying them to increase their sensitivity, resolution, speed of analysis, and range. These goals were accomplished by the design of systems that operated above ambient temperatures and at high pressures (e.g., 200 to 5000 psi). The latter was necessary because smaller exchanger particles (e.g., 10 to 20 μ) were needed to achieve good separations. These high performance systems were coupled to sensitive recording photometric units for the continuous assay of column effluents. A typical fully automated system is schematically illustrated in Fig. 7-18.

Anderson (81) pioneered the work in this field both in the concept of why special equipment was needed and in the actual designs of prototype

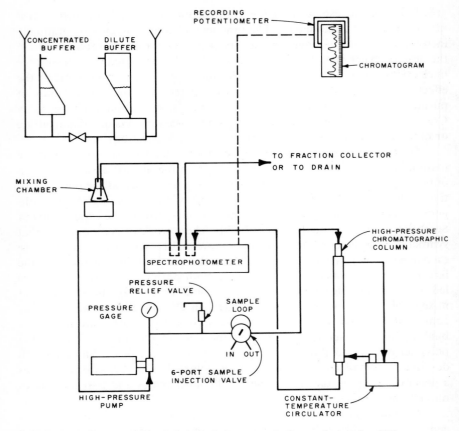

Fig. 7-18. Automated high resolution chromatograph for analyzing for UV-absorbing constituents in body fluids. (Reproduced from C. D. Scott, R. L. Jolley, W. W. Pitt, and W. F. Johnson, *Am. J. of Clin. Pathol.*, **53**, 701 (1970), by permission of the authors and J. B. Lippincott Company.)

models. Following his initial designs of rapid automated analytical systems for determining nucleotides and nucleoside derivatives, other high performance systems became available. Uziel *et al.* (82) patterned their autoanalytical nanomole-level nucleoside analyzer after the designs of Anderson (81). Other important contributions in this area have come from the efforts of Scott and his co-workers (83), who developed ion-exchange chromatographic systems for the automated high resolution analyses of the UV-absorbing constituents found in body fluids.

The speed of the separation process has been emphasized in these high performance chromatographic methods. From this standpoint, not only has thought gone into optimizing column design and operation, but also to the modification of the column packing itself. By using specially developed packing materials with a solid core and a thin porous outer shell, deep pores are eliminated, thus diminishing the path length of diffusion. Rapid mass transfer is possible in such particles since diffusion occurs only in a thin layer on the surface. Ion exchangers with such properties have been prepared. They consist of a superficially porous crust of an ion exchanger surrounding an impervious, spherical silica core. Such particles are called *pellicular* ion exchangers. These newer ion-exchange materials have been used by Kirkland (84) for the high speed separation of nucleic acid bases and other nucleic acid derivatives. Kirkland (85) has written a general review describing these materials as well as describing the present state-of-the-art of columns designed for high speed, high performance liquid chromatography.

Separation procedures for various nucleic acid derivatives are given in the remainder of this chapter. Most of the examples are taken from publications that emphasize the current trends in nucleic acid ion-exchange methodology. However, to illustrate a point or elaborate on subject matter, some examples of separations from the older literature are compared to those in more recent publications.

2. Purine and Pyrimidine Bases and Nucleosides

The separation of purine and pyrimidine bases on the NH_4-form of a strong-acid cation exchanger is shown in Fig. 7-19 and on the H-form of this exchanger in Fig. 7-20. In the former case, gradient elution is carried out at pH 4 with increasing concentrations of ammonium formate, and in the

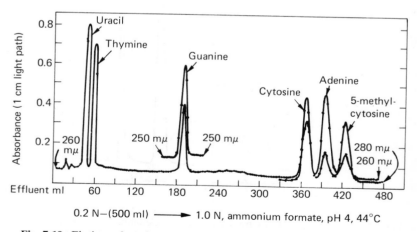

Fig. 7-19. Elution of purines and pyrimidines from 0.9×60 cm column of NH_4-form Dowex 50-X4, 200–400 mesh. Elution by gradient indicated on figure at about 7 ml/hr. Each principal peak contained about 60 μg of base. (Reproduced from C. F. Crampton, F. R. Frankel, A. M. Benson, and A. Wade, *Anal. Biochem.* **1**, 249 (1960), by permission of the authors and Academic Press, Inc.)

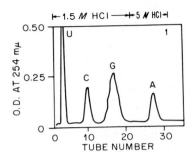

Fig. 7-20. Separation of purines and pyrimidines by cation exchange. Column 0.28 sq cm × 18 cm Dowex 50, H-form. Elution as indicated at 0.5 ml/min. Sorbed material: 3.2 O.D. units of uracil (U), 2.5 O.D. unit of cytosine (C), 2.8 O.D. units of guanine (G), and 2.0 O.D. units of adenine (A) as measured at 254 nm in a cell of 1-cm light path. (Reproduced from J. X. Khym and M. Uziel, *Biochemistry.*, **7**, 422 (1968), by permission of the authors and the American Chemical Society.)

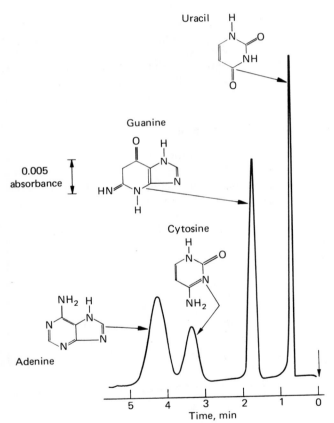

Fig. 7-21. Separation of RNA nucleic acid bases. Column, 1 m × 2.1 mm, i.d., "Zipax" strong cation exchanger, 20–37 μ; column temperature, 63°C; carrier, 0.01 N HNO$_3$; flow-rate, 2.00 cc/min; column input pressure, 735 psi; sample, 1.5 μl of 0.25 mg/ml each in 0.01 N HNO$_3$. (Reproduced from J. J. Kirkland, *J. Chromatog. Sci.*, **8**, 72 (1970), by permission of the author and The Preston Technical Abstract Co.)

latter case the separation is effected with stepwise additions of hydrochloric acid solutions. Although not directly comparable, either of these systems is applicable to the analysis of simple mixtures of purine and pyrimidine bases that contain about 50 μg of each base. These analyses were performed manually with simple, modest types of equipment. With more sophisticated equipment, nanomole quantities of purines and pyrimidines may be separated by high speed chromatography on the pellicular resin "Zipax" (duPont's trademark for controlled surface porosity supports). The porous crust of the exchanger is a fluoropolymer containing sulfonic acid groups (84) and, as seen in Fig. 7-21, a mixture of adenine, cytosine, guanine, and uracil may by separated in about five minutes on this cation exchanger.

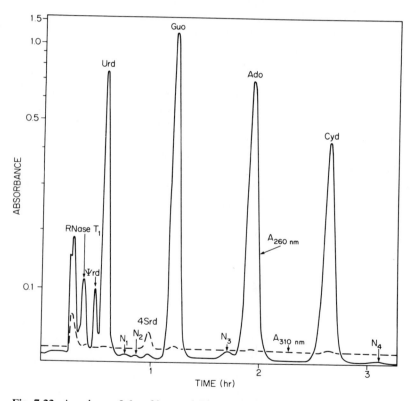

Fig. 7-22. A column 0.6 × 23 cm of Bio-Rad-A-6 was used to separate the nucleosides obtained from mixed *E. coli* B tRNA. About 3 nmoles of tRNA were analyzed. ψ rd = pseudouridine; Urd = uridine; 4 Srd = 4-thiouridine; Gua = guanosine; Ado = adenosine; Cyd = cytidine. Elution at 0.25 ml/min with 0.4 *M* ammonium formate, pH 4.65, 48°C. (Reproduced from M. Uziel and C. Koh, *J. Chromatog.* **59**, 188 (1971), by permission of the authors and Elsevier Publishing Company.)

There are certain practical advantages to be gained, in nucleic acid structural studies, by degrading polynucleotides enzymatically to the nucleoside level rather than chemically arriving at a mixture of either the bases or nucleotides (82). Accordingly, Uziel *et al.* (82) devised a moderately fast automated system that assays nucleosides by cation-exchange chromatography. As seen in Fig. 7-22, this single-solvent system will easily distinguish the major nucleosides at the nanomole level in a few hours. In addition, more complicated mixtures may be analyzed by this cation-exchange procedure, including purine and pyrimidine bases, if they are present in the mixtures. A similar system has been developed by Singhal and Cohn (86) for the analysis of nucleoside materials by anion-exchange chromatography.

Nucleosides are also separable on pellicular resins. Horvath and Lipsky (87) demonstrated that 0.03 to 10 nanomoles of nucleosides are separated on sulfonated divinylbenzene-coated glass beads in less than five minutes. Such a separation is shown in Fig. 7-23. This system may be modified to determine nucleosides at the picomole level.

More commonly, nucleotides are analyzed as single components distinguished from mixtures of nucleosides and the purine and pyrimidine bases. However, such multiple mixtures may be analyzed by a single chromatographic procedure. In this method Murakami *et al.* (88) employed an eluent of fixed ionic strength to elute nucleotides in the first part of the chromatogram, then used a four-component gradient system to elute the remaining components from a column of Aminex A-4 (Bio-Rad). In this manner, complex mixtures containing nucleotides, nucleosides, purines, and pyrimidines are analyzed.

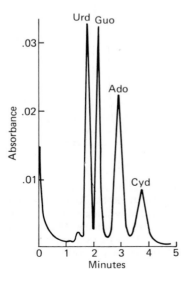

Fig. 7-23. Rapid separation of ribonucleosides. Column: 151.7 cm long; 1-mm i.d.; packed with pellicular cation exchange resin, sieve fraction #270/325 Eluent: 0.02 M $NH_4H_2PO_4$, pH 5.6. Temperature, 39°C. Flow rate, 25.5 ml/h. Inlet pressure, 131 atm. Sample size: 300 picomoles of each component. Full scale 0.04 A.U. (Reproduced from C. G. Horvath and S. R. Lipsky, *Anal. Chem.*, **41**, 1227 (1969), by permission of the authors and the American Chemical Society.)

3. Nucleoside Phosphates

Simple mixtures of mononucleotides at the micromole level may be separated in the conventional manner on strong-base anion columns (73). Usually a short column (e.g., 1 sq. cm $\times$ 10 cm of Dowex 1-Cl) is employed and the mononucleotides are eluted with dilute hydrochloric acid solutions. Manual quantitation of peaks may be made by ultraviolet spectrophotometry by which aliquots of the fractions are successively transferred from the collection tubes to quartz absorption cells. Mixtures of deoxy- or ribomononucleotides, including isomers among the latter, may be separated by these techniques. Further details are found in Cohn's (73) original papers and in his reviews (89) of this subject.

A more unusual way to separate simple mixtures of deoxy- or ribo-

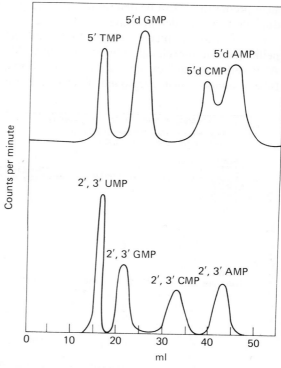

Fig. 7-24. Separation of nucleotides on a cation exchanger. Column, 0.8 $\times$ 85 cm of Bio-Rad AG 50W-X4, *Ca.* 400 mesh, NH_4-form. Elution at 1 ml/min with 0.25 *M* ammonium formate adjusted to pH 4.1 with formic acid. Upper tracing 5′-mononucleotides; lower tracing, 2′- and 3′-mononucleotides. (Reproduced from F. R. Blattner and H. P. Erickson, *Anal. Biochem.*, **18**, 220 (1967), by permission of the authors and Academic Press, Inc.)

mononucleotides is by chromatography on cation-exchange columns. Blatt-ner and Erickson (90) demonstrated the feasibility of doing this, as is shown by the chromatogram of Fig. 7-24. This separation is carried out at room temperature on a strong-acid cation exchanger with ammonium formate buffers of pH 4 as the eluent. Whether this is cation-exchange chromato-graphy or the negatively charged nucleotides are separated by the type of chromatography as discussed in Chapter 8 remains to be resolved. Blattner and Erickson (90) suggest that despite the net negative charge on the nu-cleotides, the positively charged base portion of the nucleotides (UMP and TMP compounds excluded) can interact with the negative groups of the cation resin, thus giving rise to a "form" of cation exchange.

Often, the question of purity arises in biochemical work. For example, is a certain stock of adenosine-5′-di- or triphosphate of good quality? The amount of adenine, adenosine, or adenosine-5′-monophosphate (AMP) in such preparations can be determined by a very simple anion-exchange proce-dure (91) that is demonstrated in Fig. 7-25. This same analysis has been carried out at the picomole level by Shmukler (92) utilizing high pressure pellicular anion-exchange chromatography. Mixtures of AMP, ADP, and ATP were separated in about 25 minutes. Although not directly related to the type of analyses just mentioned, Warner and Finamore (93) have originated

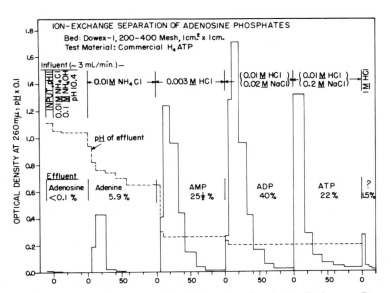

Fig. 7-25. Ion-exchange separation of adenosine phosphates. Exchanger: Dowex 1-chloride, 200–400 mesh, 1 sq cm × 1 cm. Abscissa: ml through column. For analysis, 1–10 mg of ADP or ATP in about 2 ml of 1 M NH₄OH was sorbed to the column. Elution as indicated at 3 ml/min. (Reproduced from W. E. Cohn and C. E. Carter, *J. Am. Chem. Soc.*, **72**, 4273 (1950), by permission of the authors and the American Chemical Society.)

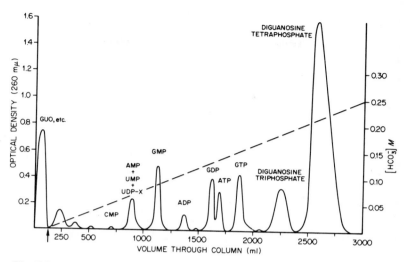

Fig. 7-26. Fractionation of guanosine derivatives. A deacidified extract containing acid-soluble nucleotides was made 0.002 M with respect to NH$_4$HCO$_3$, pH 8.6, and applied to DEAE-cellulose (HCO$_3$⁻-form) column, 1 × 50 cm. The column was washed with 0.002 M NH$_4$HCO$_3$, pH 8.6, and the nucleotides were eluted with a linear gradient of NH$_4$HCO$_3$, pH 8.6, as indicated. Recovery of ultraviolet-absorbing material was always greater than 95%. (Reproduced from A. H. Warner and F. J. Finamore, *J. Biol. Chem.*, **242**, 1933 (1967), by permission of the authors and the American Society of Biological Chemists, Inc.)

an anion-exchange procedure to distinguish guanosine derivatives from each other. Furthermore, these guanosine compounds are separated from similar substances such as AMP, ADP, and ATP when present. As seen in Fig. 7-26, increasing concentrations of ammonium bicarbonate solution were used to separate the guanine-containing substances on DEAE-cellulose columns.

Dinucleoside monophosphates are one of the smallest units of ribonucleic acids and these units are used for studying the specificity of different ribonuclease enzymes. Lapidot and Borzilay (94) have devised an anion-exchange process for the separation of 2'-5'-dinucleoside monophosphates from the corresponding 3'-5' isomers on a DEAE-Sephadex column. Such separations are demonstrated in Fig. 7-27.

Until recently, anion-exchange procedures to determine the concentration of free nucleotides in acid-extracted tissues of plant or animal origin changed very little since Hurlbert *et al.* (75) published their original procedures for such determinations. High pressure pellicular anion-exchange techniques such as those introduced by Horvath *et al.* (95) are now being used more commonly to determine the distribution of free nucleotides in tisssues. As seen in Fig. 7-28, pellicular anion-exchange resins permit the fast analysis

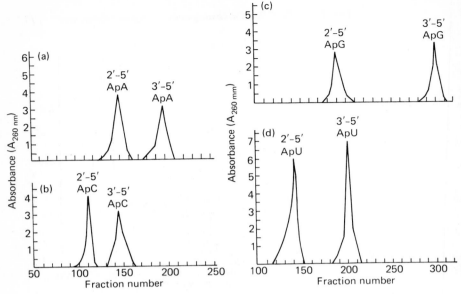

Fig. 7-27. Anion-exchange separation of isomeric dinucleoside monophosphates. Column, 3.2 × 55 cm of DEAE-Sephadex A-25 (bicarbonate form). Elution at 2.8 ml/min with a linear gradient of ammonium bicarbonate that increased in concentration from 0.02 M to 0.15 M. (Reproduced from Y. Lapidot and I. Barzilay, *J. Chromatog.*, **71**, 275 (1972), by permission of the authors and Elsevier Publishing Company.)

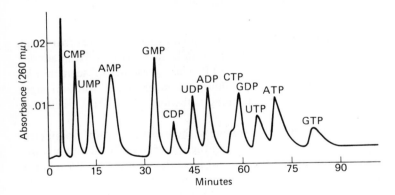

Fig. 7-28. Separation of ribonucleosides mono-, di-, and triphosphoric acids. Column: 1.0-mm i.d., 193-cm length; packed with pellicular anion exchanger (strong basic), sieve fraction No. 270–325; 8 micro-equivalent per gram. Eluent: linear gradient of ammonium formate buffer, pH 4.35 at 25°C, from 0.04 M to 1.5 M. Temperature: 71°C. Flow rate: 12 ml per hour. Flow velocity: 1 cm second^{-1}. Pressure: 51 atm. Sample: 1.5–3.5 nanomoles (0.6–1.5 μg) per component. (Reproduced from C. G. Horvath, B. A. Preiss and S. R. Lipsky, *Anal. Chem.*, **39**, 1422 (1967), by permission of the authors and the American Chemical Society.)

of minute quantities of nucleoside 5'-mono-, di-, and triphosphoric acids. Deoxymononucleotides and the 2'- and 3'-ribomononucleotides can also be separated by pellicular ion-exchange chromatography (96). Brown (97) has used these techniques extensively to study how metabolic processes are affected by drugs.

4. Polynucleotides

Homologous oligonucleotides are separated on cellulosic exchangers according to their chain length. A good example of this can be seen in the chromatogram of Fig. 7-29 taken from the work of Bollum *et al.* (98) who separated the oligodeoxyadenylates, $(dA)_2$ to $(dA)_{11}$. Peak 1 $(dA)_2$ through peak 10 $(dA)_{11}$ of Fig. 7-29 all had adenylate spectra and analyses for total and terminal phosphate gave ratios of approximately 2, 3, 4, etc., respectively. Thus the respective peaks in Fig. 7-29 are the homologous series of di- to undecanucleotides.

Larger mixed polynucleotides containing both purine and pyrimidine bases are more difficult to separate apparently due to hydrogen binding between the polynucleotides and the cellulosic exchanger. Tomlinson and Tenner (79) have overcome this difficulty by carrying out the chromatography with 7 *M* urea added to eluents. This allows separation to take place on the basis of the net negative charge of the nucleoside polymers. However, Bartos *et al.* (99) have shown that such separations occur only if the oligonucleotides in the same peak have the same purine-pyrimidine ratios, the absolute base sequence in the polymers not being of importance. If the ratio of purine to pyrimidine is different in different oligomers containing the same number of residues, fractionation within such a group may occur. Thus in these cases the ratio of purine to pyrimidine, in addition to net charge, affects the chromatography of oligonucleotides on DEAE-cellulose in the presence of 7 *M* urea.

The separation of oligonucleotides and polynucleotides is of particular

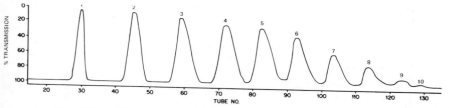

Fig. 7-29. Chromatography of homologous polydeoxyadenylates. A linear gradient of 0.02 *M* Na acetate, pH 4.8 (1 liter) to 0.4 *M* NaCl, 0.02 *M* Na acetate, pH 4.8 (1 liter) was used to elute a 1 × 25 cm DEAE-cellulose column, acetate form. (Reproduced from F. J. Bollum, E. Groeniger, and M. Yoneda, *Proc. Nat. Acad. Sci. USA.*, **51**, 853 (1964), by permission of the authors and the American Chemical Society.)

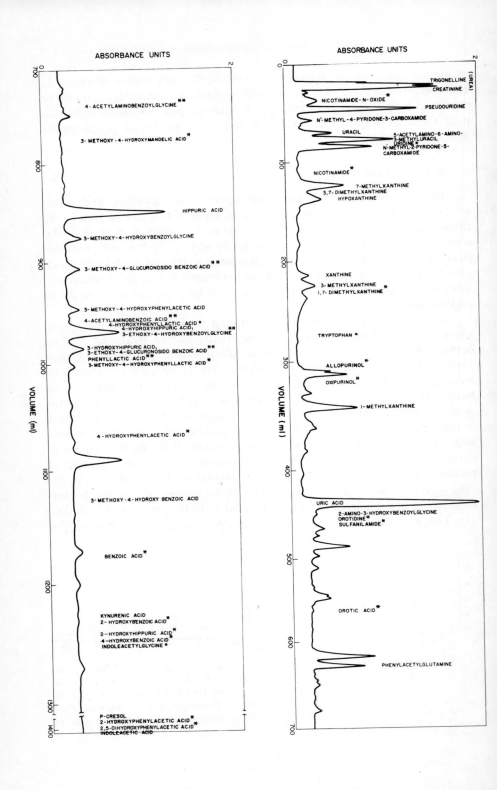

importance in the determination of base sequence in nucleic acid work, both from the standpoint of the degradation of naturally occurring nucleoside phosphate polymers and of those synthesized by chemical or enzymatic means. Various parameters associated with the cellulose column chromatography of oligo- and polynucleotides are discussed by Staehelin (80). The reader is directed to this review for refinements of this type of ion-exchange chromatography.

5. Nucleic Acids

The chromatography of nucleic acids is a very specialized field. There are as many DNA's and RNA's as there are different organisms and in turn each individual organism is a source of a wide variety of polynucleotide substances. Researchers in this field have confronted this problem by inventing almost as many chromatographic materials as there are nucleic acids. No attempt is made here to evaluate these chromatographic materials in relation to their use in nucleic acid work. This has been done in recently written comprehensive reviews on this subject, both for the fractionation of RNA's and DNA's. The former topic is covered in a lengthy (three parts), detailed article by Kothari and Taylor (100) who discuss the fractionation of RNA on modified celluloses. The literature surveyed for this article is extensive. In a similarly prepared article, Kothari (101) reviews present chromatographic methods of fractionating DNA.

H. SIMULTANEOUS ANALYSIS OF DIFFERENT CLASSES OF COMPOUNDS

A trend in the modern practice of liquid chromatography is to examine different classes of compounds with one chromatographic system. Constituents in the effluent may be determined by single monitoring devices or by such devices used in parallel, such as a system for the simultaneous determination of ultraviolet-absorbing compounds and ninhydrin-positive compounds.

Fig. 7-30. Anion-exchange chromatogram of a composite normal urine sample showing identified peaks. Compounds marked with * were identified in pathologic urine samples while compounds marked with ** were identified in a sample from a normal individual on a chemically defined synthetic diet, and positions superimposed on the normal urine chromatogram. Column, 0.62×150 cm of Aminex A-27 (Bio-Rad), 12–15 μ beads. Elution at 90 ml/hr with ammonium acetate-acetic acid buffer solution of pH 4.3 increasing in acetate concentration from 0.015 M to 6.0 M. Temperature program, ambient increasing to 60°C after 12 hr. (Reproduced from J. E. Mrochek, W. C. Butts, W. T. Rainey, Jr., and C. A. Burtis, *Clin. Chem.*, **17**, 72 (1971), by permission of the authors and the American Association of Clinical Chemists.)

Scott and his co-workers (83) introduced this type of ion-exchange methodology into the field of clinical chemistry. The purpose of this is to examine complex biological samples such as body fluids. Once the concentration of normal constituents in these samples has been determined, such values may be compared with the ion-exchange chromatographs of patients having various metabolic or pathologic abnormalities. In this manner, correlation between the normal and disease states may be made. The initial approach in such studies is to identify the normal constituents of a body fluid sample. Figure 7-30 shows the profiles obtained by the anion-exchange chromatography of a composite urine sample. Elution positions of the individual peaks were fixed by continuous ultraviolet monitoring. Mrochek *et al.* (102) utilized analytical techniques such as gas chromatography and mass spectral analyses to identify many of the eluate peaks of Fig. 7-30.

Stevenson and Burtis (103) analyzed the common constituents of analgesic tablets by anion-exchange chromatography. In this case a pellicular anion-exchange column (0.1 × 300 cm) was employed with 1.0 M tris (hydroxymethyl)-aminoethane buffer of pH 9.0 as the single eluent. By this procedure, microgram amounts of caffeine, aspirin (acetylsalicylic acid), 4-hydroxyacetanilide, phenacetin (4-ethoxyacetanilide), and salicylamide were cleanly separated in the order given in about 22 minutes. The aromatic substances were monitored with an ultraviolet absorption detector.

Finally, a unique application of simultaneous analysis is demonstrated by the work of Bonnelycke *et al.* (104). They devised an automated chromatographic system to determine mixtures of purines and pyrimidines and amino acids. This analytical scheme was developed for a miniaturized automatic analyzer for *in situ* analyses of lunar and Martian soil samples. These workers modified an amino acid analyzer so it would detect ultraviolet-absorbing compounds as well as ninhydrin-positive substances. By this type of instrumentation, mixtures containing as many as 32 purines and pyrimidines and 19 amino acids may be analyzed.

Part II

Ion-Exchange Chromatography of Inorganic Substances

A. EMERGENCE OF ION-EXCHANGE ELUTION CHROMATOGRAPHY

The high resolving power of ion-exchange chromatography was first demonstrated in work originated at Oak Ridge, Tennessee as part of the Manhattan Project during World War II. This came about by the development of ion-exchange procedures for the separation of the rare earths and other fission products. The original (72) and related and ensuing investigations by many contributors have been well documented (105, see also any general textbook on ion exchange) so that details of this work need not be given here. However, the key (that of complex-ion formation) that led to the successful, sharp separations of the closely related rare earths is discussed in some detail, since complex-ion formation is the mechanism many investigators employ to study ion-exchange metal interactions.

Elution of sorbed solutes from an ion-exchange column utilizes conditions that decrease the affinities of these bound counter-ions, releasing them from the exchanger. In the case of simple inorganic substances, elution can be achieved by adjusting the ionic strength so as to displace the sorbed counter-ions by mass action, or by lowering the charge of the sorbed ion by a change of pH, or by complexation, or by both of the latter processes. Complex-ion formation and control of pH was the key to sharp differentiations among the rare earths by ion-exchange processes. Differentiation by elution at high ionic strength was impractical since the trivalent rare earths all exhibit essentially the same strong affinity for a cation exchanger due to their almost identical chemical properties.

The technique of changing the charge of a species by complex-ion formation and thus modifying its attraction to an exchanger has the advantage that this may happen to a varying degree for each species competing for the complexing reagent. Also the displacement of the complex-ion equilibrium is in most cases pH dependent. Control of this parameter allows the chromatographic separation of the members of even closely related species.

It will be noted that in cation-exchange chromatography, complex-ion formation tends to remove the metal from the exchanger while in anion-exchange chromatography, if negative charges are formed, complex-ion formation increases attraction of the metal to the exchanger.

B. ION EXCHANGE–COMPLEX ION INTERACTIONS

The ion-exchange behavior of almost every metal ion, including the alkali metal ions, has been studied in the presence of complexing agents. The earlier work on the cation-exchange separation of the rare earths was carried out in the presence of citrate and tartrate, giving rise to less positive and, depending upon the pH, even negatively charged species. A variety of different organic ligands have been used by a number of different investigators, (106) to explore the effect of complex-ion formers on the ion-exchange reactions of the rare earths. An example of this type of work is seen from the studies of Dybczynski and his co-workers (107a–d) who separated the rare earths on anion exchangers in the presence of ethylenediaminetetraacetate (EDTA) and an EDTA derivative. The alkaline earths have been investigated similarly by Strelow and Weinert (108). These workers systematically explored parameters for the separation of Be, Mg, Ca, Sr, and Ba ions in the presence of the following complexing agents: acetate, formate, lactate, citrate, tartrate, α-hydroxyisobutyrate, malonate, malate, acetylacetonate, EDTA, and derivatives of EDTA.

In more general studies, Kraus and Nelson (109) made a broad survey of the anion-exchange behavior of elements in very high concentrations of hydrochloric acid. The results of these investigations have been compiled in the form of a "Periodic Table" that is given here as Fig. 7-31. It is evident from this figure that the separation of a large number of elements, either in groups or individually, is feasible by anion exchange in concentrated hydrochloric acid solutions. In work from the same laboratory, Nelson *et al.* (110) also investigated the cation-exchange behavior of elements at high ionic strength. In this study sorbabilities were measured from hydrochloric and perchloric acid solutions whose maximum concentrations were 12 *M*. As before, the data is presented in "Periodic Table" form. In addition, these workers demonstrated the separation of the alkaline earths and of some transition elements by column chromatography at high ionic strength. In a study of some 14 di- and trivalent metals, Fitz and Rettig (111) showed that these metals (Bi, Ca, Cd, Co, Cu, Fe, Ga, In, Mg, Mn, Ni, UO_2, Zn, V) have a greater affinity for a strong-base cation exchanger in acetone-water media than in aqueous media of the same hydrochloric acid concentration. These workers demonstrated the successful column separation of closely related metal species on a Dowex 50W cation exchanger by eluting with acetone-water-hydrochloric acid solutions of different compositions. Investigations of a similar nature were undertaken by Strelow *et al.* (112) who examined cation-exchange equilibrium distribution coefficients with Dowex 50W for 49 cations in nitric acid and 45 cations in sulfuric acid media. The coefficients were determined at acid concentrations from 0.1 *N* to 4.0 *N*. Some elution curves for multicomponent systems are presented to demonstrate separation possibilities. In work somewhat related, Faris and Warton (113) studied the behavior of the rare earths and related metals on anion exchangers in the presence of nitric acid-methanol mixtures. The data accumulated showed that the rare earths were eluted from a column in the order of decreasing atomic number. Those lighter than terbium had relatively large separation factors while the heavier rare earths showed little tendency to separate.

In less general studies Qureshi and his collaborators (114) examined the sorption of metal-formate complexes on cation-exchange resins. These workers present separation data for some 20 elements, among which are members of the alkaline earths, the rare earths, and the transition elements. The latter elements have also been studied in nitrite media by Bhatnagar *et al.* (115) who demonstrate the utility of transition metal-nitrite complexes in separation work. Both cation- and anion-exchange resins were used in

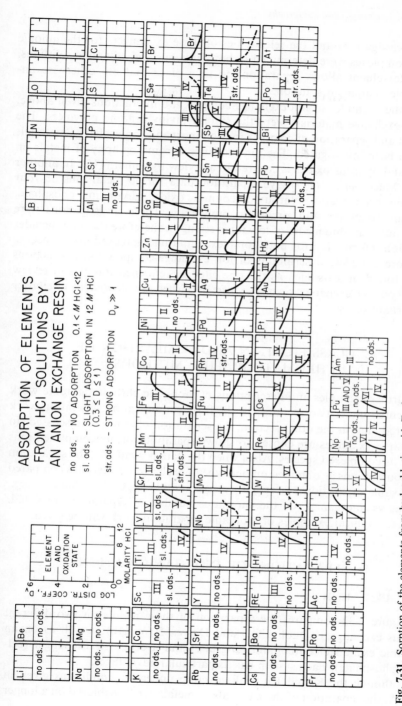

Fig. 7-31. Sorption of the elements from hydrochloric acid. Exchanger, Dowex 1-X10. As indicated in the sample plot, the data for each individual element is given as a graph of log D_v versus M HCl. (Reproduced from K. A. Kraus and F. Nelson, *Proc. 1st Intern. Conf. on Peaceful Uses of Atomic Energy*, **7**, 113 (1956), by permission of the authors and the United Nations Publications Office.)

these investigations and the selectivities of these exchangers for the anionic nitrite complexes were determined in both aqueous and ethanolic solutions. The latter eluent allows separations to take place that are otherwise unattainable in aqueous solution.

In the examples of ion exchange-complex ion interactions given above, the ion-exchange materials were the styrene-divinylbenzene polymers. The sorption and separation of complex ions on inorganic ion-exchange materials has also been investigated. The ion-exchange chromatography of metals on oxides and hydrous oxides is covered in a review by Fuller (116). In another survey, Maeck *et al.* (117) give separation data for 60 metal ions on four inorganic exchangers: hydrous zirconium oxide, zirconium phosphate, zirconium tungstate, and zirconium molybdate. These investigations were carried out in nitrate media of low ionic strength, therefore the data includes separation information for both complexed and uncomplexed metal species.

More complete descriptions of ion exchange-complex ion interactions can be found in texts such as the one cited in reference 106 and in review articles on ion exchange such as those published by *Analytical Chemistry* on alternate years.

C. CLASSIFICATION AND COMPARISON OF CHROMATOGRAPHIC METHODS

Further examples and descriptions of inorganic chromatographic work is presented under two sections. In the first section, cation separation work is covered, and in the other section, the separation of anions is discussed. In both parts, examples are chosen to show separations of noncomplexed ions as well as complexed species. Also, when the data is available, the separation of a group of ions as obtained on one type of exchanger material is compared to the same separation as carried out on another type of exchanger (e.g., resinous vs. inorganic). Of the enormous number of separations reported, only a few typical examples of inorganic separation work will be given.

D. SEPARATIONS OF INORGANIC CATIONS

1. Alkali Metals

Figure 7-32 compares the separation of these metals on two different resinous exchangers (Duolite C-3 and Dowex 50W-X16) as well as on an inorganic exchanger (zirconium tungstate) and a cellulosic exchanger (cellulose phosphate). Each system has its particular advantages.

Although the K and Rb peaks overlap under the conditions given in Panel A, the separation of these two alkali metals can be achieved on a longer column of Duolite C-3 (a sulfonated phenol-formaldehyde resin). For this separation, Nelson *et al.* (118) increased the column length to 12 cm and eluted K with 1.5 M HCl, after which Rb was eluted with 6 M HCl. If present in a mixture with other alkali ions, K and Rb are eluted as in Panel A and then rechromatographed on the longer column. Methanol is used in the eluent, mainly to effect good separation of Li from Na and of Na from K.

As seen in Panel B of Fig. 7-32, the complete separation of the alkali metals has been achieved with a single eluent on Dowex 50W-X16 (a highly crosslinked, sulfonated divinylbenzene resin). The chromatogram shown in Panel B is only one of several published by Dybczynski (119) who investigated the effect of temperature and of the degree of crosslinking of Dowex resins on the cation-exchanger separation of alkali and alkaline earth metal ions. The best separation of the alkali metals was achieved at the highest crosslinking (16%) and at ambient temperature (25°C). While Na appeared as a single peak, very poor resolution of K, Rb, and Cs was achieved on 2% crosslinked resin. Noticeable improvement in resolution was seen when 8% crosslinked resin was substituted for the 2% material. Operation of the columns at 75°C instead of 25°C gave poor resolution at all levels of crosslinking.

As illustrated in Panel C of Fig. 7-32, the affinity of the alkali metals for the exchanger, zirconium tungstate, increases sharply with atomic number (120). Each metal may be completely separated from its neighbor as a narrow solute band by elution with ammonium chloride solutions whose concentrations are sharply increased after each preceding alkali ion is removed. Separations of the alkali metal ions may also be obtained on the inorganic cation exchanger, ammonium phosphomolybdate, as reported by Coetzee and Rohwer (121) and on hydrous tin(IV) oxide as demonstrated by Donaldson and Fuller (122).

The separation of the alkali metal ions is easily performed on cellulose phosphate (123) as demonstrated in Panel D of Fig. 7-32. Widely separated peaks are obtained with a single dilute eluent, 0.1 M hydrochloric acid. The time required to completely separate Na, K, Rb, and Cs under the conditions given for Panel D is about eight hours. However, with a mixture of 80% methanol-2 M hydrochloric acid, a Na-K separation is obtained on a column (0.6 × 8.0 cm) of cellulose phosphate within one hour (Na is eluted in the first 30 ml, K in the 50–80 ml fraction).

2. Alkaline Earth Ions

Several different chromatographic procedures are available for the separation of this group of ions. Four such choices are compared in Fig. 7-33. Among these, two procedures demonstrate the separation of non-complexed species, while in the other two methods elution of the alkaline earths is carried out in the presence of complex-ion formers.

Panel A

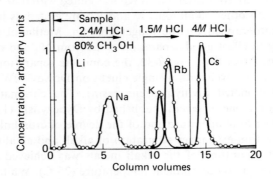

Exchanger: Duolite C-3, 0.28 cm² × cm.
Elution with eluents as shown at 0.4 cm/min.

Panel C

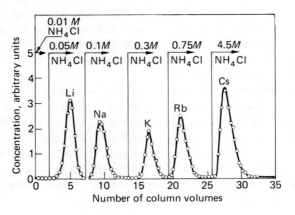

Exchanger: Zirconium tungstate, 0.13 cm² × 12.3 cm.
Elution with eluents as shown at 0.75 cm/min.

Fig. 7-32. Comparison of exchanger types for the cation-exchange chromato-graphy of the alkali metal ions: Panel A (Reproduced from F. Nelson, D. C. Michelson, H. O. Phillips, and K. A. Kraus, *J. Chromatog.*, **20**, 107 (1965), by permission of the authors and Elsevier Publishing Co.); Panel B (Reproduced from R. Dybczynski, *J. Chromatog.*, **71**, 507 (1972), by permission of the author and Elsevier Publishing Co.), Panel C (Reproduced from K. A. Kraus, H. O.

188

Panel B

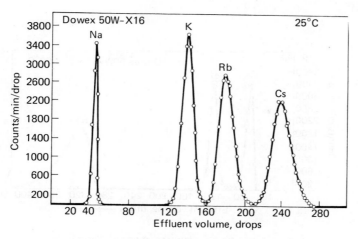

Exchanger: Dowex 50W-X16, 0.031 cm² × 7 cm.
Elution with 0.636 *M* HCl at 0.72 cm/min.

Panel D

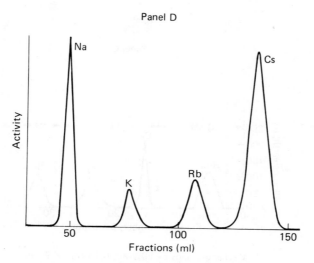

Exchanger: Cellulose phosphate 0.50 cm² × 30 cm.
Elution with 0.1 *M* HCl at 0.7 ml/min.

Phillips, T. A. Carlson, and J. S. Johnson, *Proc. 2nd Intern. Conf. on Peaceful Uses of Atomic Energy*, **28**, 3 (1958), by permission of the authors and the United Nations Publications Office); Panel D (Reproduced from J. G. van Raaphort and H. H. Haremaker, *Talanta*, **17**, 345 (1970), by permission of the authors and Pergamon Press).

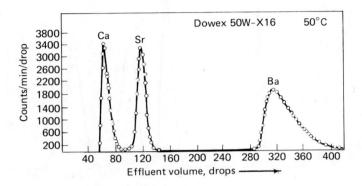

Exchanger: Dowex 50W-X16, 0.031 cm² × 7 cm.
Elution with 2.04 *N* HCl at 0.74 cm/min.

Panel C

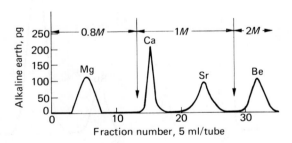

Exchanger: Dowex 50W-X8, 0.13 cm² × 9 cm.
Elution with ammonium α-hydroxyisobutyrate
solutions as shown at 0.65 ml/min.

Fig. 7-33. Comparison of four different chromatographic procedures for the separation of the alkaline earths: Panel A (Reproduced from R. Dybczynski, *J. Chromatog.*, **71**, 507 (1972), by permission of the author and Elsevier Publishing Co.); Panel B (Reproduced from K. A. Kraus, H. O. Phillips, T. A. Carlson, and J. S. Johnson, *Proc. 2nd Intern. Conf. on Peaceful Uses of Atomic Energy*, **28**, 3 (1958), by permission of the authors and the United Nations Publications

Panel B

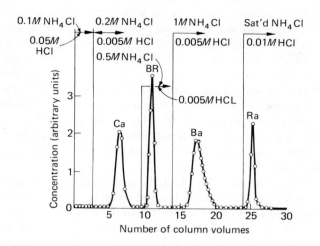

Exchanger: Zirconuim molybdate, 0.19 cm² × 10.0 cm.
Elution with eluents as shown at 1.1 cm/min.

Panel D

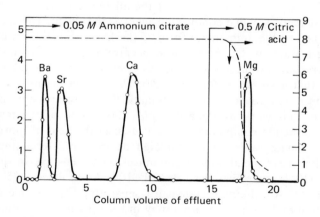

Exchanger: Dowex-1, 0.27 cm² × 44 cm. Elution
with eluents as shown at 0.4 cm/min.

Office); Panel C (Reproduced from F. H. Pollard, G. Nickless, and D. Spincer, *J. Chromatog.*, **13**, 224 (1964), by permission of the authors and Elsevier Publishing Co.); Panel D (Reproduced from F. Nelson and K. A. Kraus, *J. Am. Chem. Soc.*, **77**, 801 (1955), by permission of the authors and the American Chemical Society).

Panel A of Fig. 7-33 shows the separation of the alkaline earths on a 16% crosslinked Dowex cation-exchange resin, with hydrochloric acid as the eluent (119). As the temperature is decreased from 50°C, poorer resolution is obtained among the three elements of Panel A (Be, Mg, and Ra were not included in these studies). With an 8% crosslinked resin, the most efficient separation of the Sr from Ba is obtained, but poorer resolution of the Ca from Sr results. When the separation is carried out on resins of lower crosslinkage ($<$ 8%), the resolution becomes very poor. Strelow and Weinert (108) give distribution coefficient data for Be, Mg, Ca, Sr, and Ba to indicate that complete separation of alkaline earth mixtures by cation-exchange chromatography is possible when various organic complexing agents are added to eluents (e.g., see Panel C).

As demonstrated in Panel B of Fig. 7-33, the elements calcium to radium may be completely separated from each other on a small column of zirconium molybdate (120). Because of the large differences in the affinities of the alkaline earth ions towards this exchanger, narrow elution bands are obtained only when stepwise or gradient elution techniques are utilized. To achieve the sharp bands shown in Panel B, the ionic strength of the eluent was increased markedly after each element emerged. Particularly noticeable is the large difference in the selectivity of the exchanger for barium as compared to radium. These elements can be separated with columns only a few centimeters in length.

In Panel C is shown the separation of the alkaline earths on a cation exchanger (Dowex 50W), with a complexing agent as the eluent. The alkaline earths were sorbed to the ammonium form of the Dowex resin from dilute acid; following a water wash, increasing concentrations of ammonium α-hydroxyisobutyric buffer solutions were used as eluents, which gave rise to the elution pattern that is shown in Panel C (124). The alkaline earths emerging from the column were assayed with an automatic flame spectrophotometric detection device. This work by Pollard *et al.* (124) is one of the few instances where automated equipment has been used in the field of inorganic ion-exchange chromatography. An automated device for determining the alkali and alkaline earth metal ions in column effluents sensitive in the ultramicro range (10^{-12} mole) has been described by Araki *et al.* (125) in a more recent report. An alternate system for separating the alkaline earth elements by cation-exchange chromatography in the presence of a complexion former was devised by Milton and Grumitt (126). These workers separated Mg, Ca, Sr, Ba, and Ra with 1.5 *M* ammonium lactate at pH 7 on a column (8 × 1.1 cm) of Dowex 50.

The separation of the alkaline earths by anion-exchange chromatography is demonstrated in Panel D of Fig. 7-33. In this case the anionic citrate complexes of the alkaline earths were separated on the citrate form of the strong-base anion exchanger, Dowex 1 (127). As seen in Panel C, the degree to which the alkaline earths are complexed is sufficiently different at a given concentration of ammonium citrate so that each negative species has a different sorbability. These differences are the basis for the chromatographic separation of the alkaline earth citrate complexes on anion-exchange columns.

3. Transition Elements and Related Metals

Three different modes for the chromatographic separation of some heavy metals are compared. The first such system utilizes the tartrate form of Dowex 2-X8 to separate the tartrate complexes of some transition metals, as is demonstrated in Fig. 7-34 (128). The basis for this separation is analogous to that given for the citrate complexes of the alkaline earths (see Panel D of Fig. 7-33). Control of the tartrate ion concentration and pH are the main parameters affecting the degree of resolution of the transition elements. In addition to the metal separation shown in Fig. 7-34, mixtures of Cr, Ni, Fe, Mo, or Mn, Cr, Fe, Mo, or Zn, Cd, Hg, may be resolved as their tartrate complexes by this anion-exchange procedure of Morie and Sweet (128).

Trace transition metal ion impurities in aluminum and lead samples have been determined by Bishay (129) by an anion-exchange procedure. This,

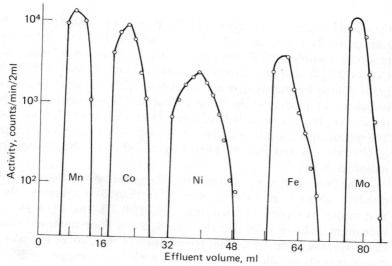

Fig. 7-34. Separation of the transition elements as their tartrate complexes by anion-exchange chromatography. Conditions: column, 0.8×8.0 cm Dowex 2-X8, tartrate form; elution with 8.5×10^{-2} tartrate solution at pH 4.0 for Mn, Co, and Ni, 0.1 M HCl for Fe, and 3 M NaOH for Mo at 3 drops/min. (Reproduced from G. P. Morie and T. R. Sweet, *J. Chromatog.* **16**, 201 (1964), by permission of the authors and Elsevier Publishing Co.)

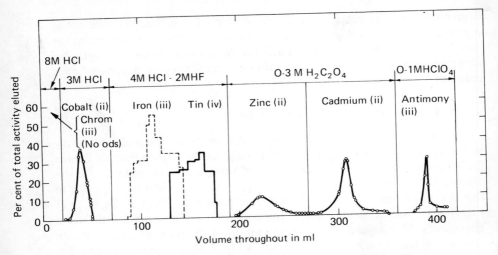

Fig. 7-35. Separation of some transition elements by anion-exchange chromatography with acidic eluents. Conditions: column, 1.2 × 25 cm Amberlite CG 400-X8, form changes with each new eluent; elution with eluents as shown at 0.5 ml/min. (Reproduced from T. Z. Bishay, *Anal. Chem.*, **44**, 1087 (1972), by permission of the author and the American Chemical Society.)

the second mode, also depends upon the presence of complex-ion formers in eluents to effect separation. But as seen in Fig. 7-35 the chromatographic process is complicated. Samples are sorbed to the anion column (Amberlite CG 400-X8) from 8 M hydrochloric acid solution. As seen in the chromatogram, among the metals separated, chromium does not form a strong chlorocomplex and appears in the sample effluent. The negative cobalt species is more strongly attracted to the exchanger but is eluted with 3 M acid solution. The remaining metal ions are all strongly held to the exchanger in the presence of chloride ion, but may be selectively desorbed by the eluting agents as indicated in Fig. 7-35.

The third mode concerns the ion-exchange behavior of metals on DEAE-cellulose columns in the presence of methanol-thiocyanate-hydrochloric acid media, as reported by Kuroda *et al.* (130). The weakly basic cellulosic exchanger sorbs very few metals from aqueous thiocyanate-chloride media in the absence of methanol. But the addition of the alcohol to make a three-component eluent allows many heavy metals to be sorbed on DEAE. The separation of the thiocyanato-complexes of cadmium, copper, zinc, and bismuth on a column of DEAE-cellulose is illustrated in Fig. 7-36. As seen in the figure, the eluent composition is readjusted after each transition metal ion is eluted from the column.

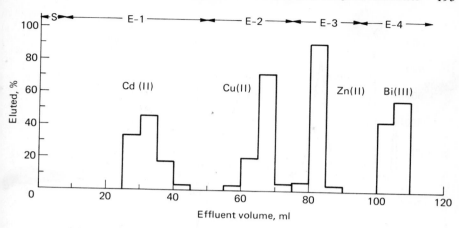

Fig. 7-36. Separation of the thiocyanato complexes of cadium, copper, zinc, and bismuth. Conditions: column, 0.13 × 4.5 cm DEAE-cellulose, thiocyanate form; elution as shown at 1 ml/min. Eluent, E-1: 1.1 M NH$_4$ SCN—0.0011 M HCl in 95% methanol; E-2: 1.1 M HCl in 80% methanol; E-3: 1.1 M HCl in 50% methanol; E-4: 2.1 M HCl; sample S, sorbed from E-1 eluent. (Reproduced from R. Kurodo, T. Kondo, and K. Oguma, *Talanta*, **19**, 1043 (1972), by permission of the authors and Pergamon Press.)

4. The Rare Earth Elements

These closely related elements form strong complexes, mostly of negative charge, with several organic ligands. As pointed out previously in earlier publications, several different groups of workers have investigated the separation of the rare earths on cation-exchange resins with the aid of complexing agents. However, the behavior of the rare earths on anion-exchange resins had been little investigated until the thorough studies of Dybczynski and his co-workers (107a–d). These workers chose the multidentate organic ligand EDTA and a derivative of it, trans-1,2-diaminocyclohexane-N,N-tetraacetic acid (DCTA) as complexing agents. Strongly basic anion exchangers such as the Dowex or Amberlite resins were selected as the ion exchangers.

In studies with EDTA (107a), it was found the affinity to the ion exchanger increased in a regular manner within the cerium group, reaching a maximum at europium, and then decreased on going to higher atomic numbers. The charge of the complex ions was -1 for all lanthanides and yttrium at pH's within the range 4.55 to 4.70. In this lanthanide group, the observed large differences in distribution coefficients (at pH's 4.55 to 4.70) may possibly be related to the differences in the hydration of the complex ions of this group. The distribution coefficients obtained in the presence of EDTA for neigh-

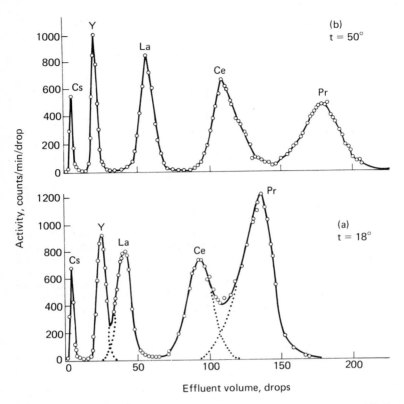

Fig. 7-37. Influence of temperature on the separation of Y, La, Ce, and Pr in the presence of EDTA. Conditions: column, 0.036 cm² × 4.2 cm Amberlite IRA-400, EDTA-form; Elution with 0.0447 EDTA (pH 4.5–4.7) at 1.12 cm/min; Temperature, panel (a) 18°C and panel (b) 50°C. (Reproduced from R. Dybczn-ski, *J. Chromatog.*, **14**, 79 (1964), by permission of the author and Elsevier Publishing Co.).

boring rare earth elements indicated the possibility of achieving separations on short columns. Such a separation is shown in Fig. 7-37 for the Y, La, Ce, and Pr group (107b). This figure demonstrates that an increase of temperature increases the resolution between the elements shown. However, in other cases (e.g., elution of the pairs Ce-Eu and Sc-Ho) an increase in temperature results in a decrease in resolution (107b).

Investigations with DCTA revealed the same unusual affinity of rare earths towards strongly basic anion exchangers that was shown with EDTA (107c). Selectivity of the exchangers increased in a regular manner from lanthanum to promethium, then decreased with further increase of atomic number. Other observations were that selectivity coefficients, separation factors, resolution, and especially plate heights depended markedly on resin crosslinking (107d). For example, the best separation of the DCTA complexes of Tm, Tb, Eu, and Pm was achieved with Dowex 1-X4. With lower or higher crosslinked Dowex resins, the resolution between pairs became poorer. In comparable instances, the plate heights determined for the DCTA system are approximately twice the corresponding plate heights for the EDTA system, indicating the ion-exchange kinetics are considerably less favorable for the rare earth-DCTA complexes.

5. Miscellaneous Metals

Tl, In, Ga, and Al are eluted in the order given from Dowex 50W-X8 columns (2.0 × 19 cm) with hydrochloric acid-acetone solutions of changing compositions (131). The affinity of these Group IIIA elements increases sharply as the atomic number decreases. Each element is widely separated from its neighbor, appearing in a sharp elution band after each eluent change. In work from the same laboratory, conditions are given for equally sharp separations for many other groups of metals. In these studies Strelow *et al.* (112) present distribution coefficient data that indicates numerous possibilities for separations of analytical interest such as the separation of Hg, Cd, Be, Fe, Ba, and Zr on a single cation-exchange column with various aqueous-acidic mixtures as the eluents.

Many groups of different metals throughout the Periodic Table have been separated from other groups and from each other on various cellulose exchangers. In most cases, the eluents for these metal separations consists of organic reagents added to aqueous solvent systems. Cellulose exchangers have been used for: the analysis of steel; the recovery and purification of gold, uranium, or thorium; the separation of precious metals such as platinum, iridium, indium, palladium, tantalum, and niobium; and for the separation of the transition elements. These separations and others are described in a detailed review by Muzzarelli (132).

E. SEPARATION OF INORGANIC ANIONS

1. Separation of Chloride, Bromide, and Iodide Ions

The quantitative analysis of mixtures of these halides has always been a difficult task. This problem has been alleviated by the development of anion-exchange procedures by which satisfactory results can be obtained even for a minor constituent of a halide mixture. Two such chromatographic procedures are compared as shown in Fig. 7-38. The separation illustrated in Panel A was carried out on the nitrate form of Dowex 1-X10 by elution with sodium nitrate solutions (133). The same eluent, but of different concentrations, was

Panel A

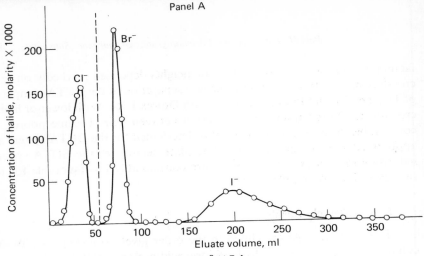

Exchanger: Dowex 1-X10, 3.7 cm² × 7.4 cm.
Elution at 1.0 ml/min. with 0.5 M NaNO$_3$ for Cl⁻, and
with 2.0 M NaNO$_3$ for Br⁻ and I⁻.

Panel B

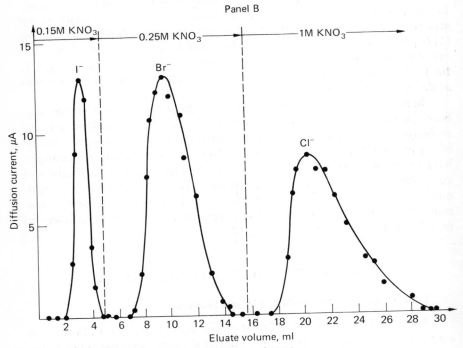

Exchanger: hydrous zirconium oxide, 0.074 cm² × 12 cm.
Elution with NaNO$_3$ solutions are shown at 0.14 ml/min.

Fig. 7-38. Comparison of exchanger types for the separation of halide mixtures:
Panel A (Reproduced from R. C. DeGeiso, W. Rieman III, and S. Lindenbaum,
Anal. Chem., **26**, 1840 (1954), by permission of the authors and the American
Chemical Society); Panel B (Reproduced from S. Tustanowski, *J. Chromatog.*,
31, 268 (1967), by permission of the author and Elsevier Publishing Co.)

also used for the separation shown in Panel B. However, in this case the exchanger consisted of hydrous zirconium oxide (134). The comparison of the two procedures is interesting in that the elution order of the halides from the zirconium oxide exchanger is just the reverse of that obtained on the resinous anion exchanger.

2. Chromatography of Polyphosphate Mixtures

Anion-exchange procedures have fulfilled a need for reasonably quick procedures for the quantitative assay of commercial phosphates and phosphate-containing products. Advances in this field stem from the original investigations of Beukenkamp *et al.* (135) who initially separated condensed phosphate mixtures on strongly basic anion exchangers by discontinuous stepwise elution techniques with increasing concentrations of potassium chloride as the eluent. In later work, Grande and Beukenkamp (136) improved the ion-exchange method for the analysis of phosphate polymers by employing gradient elution techniques. An example of this approach is seen in the separation of the phosphate mixture that is illustrated in Fig. 7-39. Lungren and Loeb (137) have automated this procedure.

The chromatographic separation of cyclic polymeric phosphates on QAE-Sephadex A-25, a quaternary-base dextran, has been achieved by Kura and Ohashi (138). Cyclic trimeta-, tetrameta-, pentameta-, and heptametaphosphates were separated from each other with 0.3 M potassium chloride

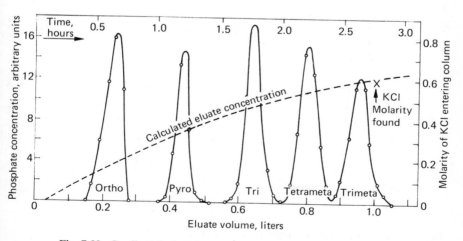

Fig. 7-39. Gradient elution chromatography of phosphate polymers. Conditions: column, 3.8 cm² × 19.0 cm Bio-Rad AG1-X8, Cl-form; elution with gradually increasing concentrations of KCl solution as shown at 6.0 ml/min; for gradient, 1 M KCl is added to a mixing vessel that initially contained one liter of water. (Reproduced from J. A. Grande and J. Beukenkamp, *Anal. Chem.*, **28**, 1497 (1956), by permission of the authors and the American Chemical Society.)

buffered at pH 5.2 with acetate and from hexameta- and octametaphosphates on a column (88 × 1.5 cm) of the Sephadex anion exchanger. The latter two cyclic phosphates were separated by elution with buffered 0.25 M potassium chloride.

3. Separation of Hypophosphite, Phosphite, and Phosphate

Iodometric methods for estimating mixtures of these phosphorus anions are unsatisfactory, especially when the mixture also contains other lower

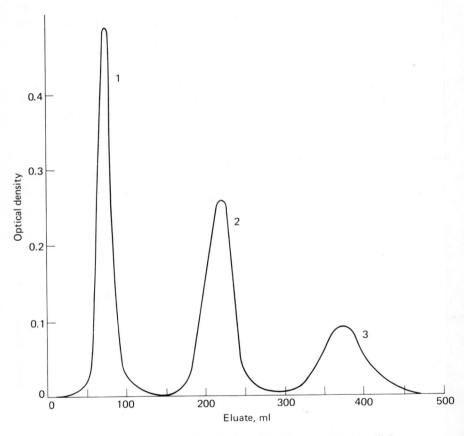

Fig. 7-40. Separation of hypophosphite (1) phosphite (2), and phosphate (3) by anion-exchange chromatography at pH 11.4 at 2°C. Conditions: column, 1.0 × 50 cm Dowex 1-X8, Cl-form; elution with a gradient of KCl solution at 25 ml/hr; gradient same as in Fig. 7-39 except that mixing vessel initially contained one liter of 0.1 M KCl. (Reproduced from F. H. Pollard, *et al.*, *J. Chromatog.*, **9**, 227 (1962), by permission of the authors and the Elsevier Publishing Company.)

oxyanions or thiooxyanions of phosphorus, since these also react with iodine. Accordingly, Pollard *et al.* (139) developed a method for separating hypophosphite, phosphite, and phosphate using gradient elution anion-exchange chromatography on Dowex-1 resins at pH 6.8 and 11.4. Such a separation at pH 11.4 is demonstrated in Fig. 7-40. The order of elution of phosphite and phosphate is reversed on changing the pH of the eluent solution from 11.4 to 6.8. This is presumably because the phosphate ion is doubly charged at the higher pH value and singly charged at pH 6.8, while the charge of the phosphite ion is the same at both pH's. In further work from the same laboratory, Pollard and his collaborators (140) devised a fully automatic chromatographic system that is capable of analyzing mixtures of lower phosphorus anions, thiophosphates, amido- and imidophosphates as well as polyphosphates.

4. Miscellaneous Other Anions

The halate ions, iodate, bromate, and chlorate, may be separated on columns (2 sq. cm × 13 cm) of Dowex 21K or Dowex 2 by stepwise elution with sodium nitrate solutions (0.5 *M* to 2.0 *M*) (141). In comparison with the halide ion separation of similar divinylbenzene-polystyrene resins, given earlier in this chapter, the order of elution is reversed. The iodate ion shows the least attraction for the anion exchangers.

Anion-exchange chromatography has also been used to separate oxygen-containing sulfur compounds such as thiosulfate, sulfite, and the lower polythionates (142). These separations were obtained on a column (0.75 sq. cm × 20 cm) of De-Acidite FF in the chloride form with increasing concentrations of hydrochloric acid solutions (3 *N* to 9 *N*) as the eluting agents.

REFERENCES

1. Moore, S., and Stein, W. H., *J. Biol. Chem.*, **192**, 663 (1951).

2a. Spackman, D. H., Stein, W. H., and Moore, S., *Anal. Chem.*, **30**, 1190 (1958).

2b. Spackman, D. H., in *Methods in Enzymology*, Academic Press, Inc., New York, 1967 (ed. by C. H. W. Hirs), Vol. XI, Article 1. Editors-in-chief S. P. Colowick and N. O. Kaplan.

3. Hamilton, P. B., *Anal. Chem.*, **35**, 2063 (1963).

4. Hamilton, P. B., in *Automation in Analytical Chemistry, Technicon Symposia* 1967, Vol. 1, Mediad, Inc., New York, 1968.

5. Piez, K., and Morris, L. A., *Anal. Biochem.*, **1**, 187 (1960).

6. Hamilton, P. B., *Nature*, **205**, 284 (1965).

7. Peters, J. H., and Berridge, B. J., *Chromatog. Rev.*, **12**, 157 (1970).

8. Saifer, A., A'Zary, E., Valenti, C., and Schneck, L., *Clin. Chem.*, **16**, 892 (1970).

9. Mardens, Y., and Sande, M. van, in *Automation in Analytical Chemistry*, *Technicon Symposia* 1967, Vol. 2, Mediad, Inc., New York, 1968.

10. Jeffsson, J. O., and Karlsson, I. M., *J. Chromatog.* **72**, 93 (1972).

11. Fowler, B., and Robins, A. J., *J. Chromatog.* **72**, 105 (1972).

12. Schroeder, W. A., in *Methods in Enzymology*, Academic Press, Inc., New York, 1972 (ed. by C. H. W. Hirs and S. N. Timasheff), Vol. XXV, Part B, Articles 15 and 16. Editors-in-chief, S. P. Colowick and N. O. Kaplan.

13. Schroeder, W. A., Jones, R. T., Cormick, J., and McCalla, K., *Anal. Chem.*, **34**, 1570 (1962); Schroeder, W. A., and Robberson, B., *Anal. Chem.*, **37**, 1583 (1965).

14. Edmunson, A. B., in *Methods in Enzymology*, Academic Press, Inc., New York, 1967 (ed. by C. H. W. Hirs), Vol. XI, Article 40. Editors-in-chief, S. P. Colowick and N. O. Kaplan.

15. Moore, S., and Stein, W. H., *Adv. Protein Chem.*, **XI**, 191 (1956).

16. Peterson, E. A., and Sober, H. A., *J. Am. Chem. Soc.*, **78**, 756 (1956).

17. Roy, D., and Konigsberg, W., in *Methods in Enzymology*, Academic Press, Inc., New York, 1972 (ed. by C. H. W. Hirs and S. N. Timasheff), Vol. XXV, Part B, Article 17. Editors-in-chief S. P. Colowick and N. O. Kaplan.

18. Peterson, E. A., *Cellulosic Ion Exchangers*, American Elsevier Publishing Company, Inc., New York, 1970 (ed. by T. S. Work and E. Work).

19. Fasold, H., Gundlach, G., and Turba, F., in *Chromatography*, Reinhold Publishing Corporation, New York, 1967 (ed. by E. Heftmann), 2nd ed., Chap. 16.

20. Clegg, J. B., Naughton, M. A., and Weatherall, D. J., *J. Mol. Biol.*, **19**, 91 (1966).

21. Popp, R. A., and Bailiff, E. G., *Biochim. Biophys. Acta*, **303**, 61 (1973).

22. Smith, M. A., and Stahman, M. A., *J. Chromatog.*, **41**, 228 (1969).

23. Nakagawa, Y., and Bender, M. L., *Biochemistry*, **9**, 259 (1970).

24. Hurwitz, J., Gold, M., and Anders, M., *J. Biol. Chem.*, **239**, 3462 (1964).

25. Khym, J. X., in *Methods in Enzymology*, Academic Press, Inc., New York, 1967 (ed. by L. Grossman and K. Moldave), Vol. XII, Part A, Article 12. Editors-in-chief S. P. Colowick and N. O. Kaplan.

26. Khym, J. X., Zill, L. P., and Cohn, W. E., in *Ion Exchangers in Organic and Biochemistry*, Interscience Publishers, Inc., New York, 1957 (ed. by C. Calmon and T. R. E. Kressman), Chap. 20.

27. Khym, J. X., and Zill, L. P., *J. Am. Chem. Soc.*, **74**, 2090 (1952).

28. Zill, L. P., Khym, J. X., and Cheniae, G. M., *J. Am. Chem. Soc.*, **75**, 1339 (1953).

29. Khym, J. X., Jolley, R L., and Scott, C. D., *Cereal Science Today*, **15**, 44 (1970).

30. Hallen, A., *Acta Chem. Scand.*, **14**, 2249 (1960).

31. (a) Walborg, Jr., E. F., Christensson, L., and Gardell, S., *Anal. Biochem.*, **13**, 177 (1965); (b) Walborg, Jr., E. F., and Lantz, R. S., *Anal. Biochem.*, **22**, 123 (1968); (c) Walborg, Jr., E. F., and Kondo, L. E., *Anal. Biochem.*, **37**, 320 (1970).

32. Kesler, R. B., *Anal. Chem.*, **39**, 1416 (1967).

33. Lee, Y. C., McKelvy, F., and Lang, D., *Anal. Biochem.*, **27**, 567 (1969).

34. Ohms, J. I., Zec, J., Benson, Jr., J. V., and Patterson, J. A., *Anal. Biochem.*, **20**, 51 (1967).

34a. Floridi, A., *J. Chromatog,.* **59**, 61 (1971).

35. Hough, L., Jones, J. V. S., and Wusteman, P., *Carbohydr. Res.*, **21**, 9 (1972).

36. Green, J. G., *Natl. Cancer Inst. Monograph No.* 21, 447 (1966).

37. Scott, C. D., Jolley, R. L., Pitt, W. W., and Johnson, W. F., *Am. J. Clin. Pathol.*, **53**, 701 (1970).

38. Carubelli, R., *Carbohydr. Res.*, **2**, 480 (1966).

39. Khym, J. X., and Cohn, W. E., *Biochim. Biophys. Acata*, **15**, 139 (1954).

40. Khym, J. X., and Cohn, W. E., *J. Am. Chem. Soc.*, **75**, 1153 (1953); Khym, J. X., Doherty, D. G., and Cohn, W. E., *J. Am. Chem. Soc.*, **76**, 5523 (1954).

41. Cohn, W. E., in *The Nucleic Acids*, Academic Press, Inc., New York, 1955 (ed. by E. Chargaff and J. N. Davidson), Vol. I, p. 238.

42. Samuelson, O., and Wictorin, L., *Carbohydr. Res.*, **4**, 139 (1967).

43. Schenker, H. H., and Rieman III, W., *Anal. Chem.*, **25**, 1637 (1953).

44. Busch, H., Hurlbert, R. B., and Potter, V. R., *J. Biol. Chem.*, **196**, 717 (1952).

45. Khym, J. X., and Doherty, D. G., *J. Am. Chem. Soc.*, **74**, 3199 (1952).

46. Johnson, S., and Samuelson, O., *Anal. Chim. Acta*, **36**, 1 (1966).

47. Samuelson, O., and Thede, L., *J. Chromatog.*, **30**, 556 (1967).

48. Carlsson, B., and Samuelson, O., *Anal. Chim. Acta*, **49**, 247 (1970).

49. Fransson, L. A., Roden, L., and Spach, M. L., *Anal. Biochem.*, **21**, 317 (1968).

50. Lee, K. S., and Samuelson, O., *Anal. Chim. Acta*, **37**, 359 (1967).

51. Bengtsson, L., and Samuelson, O., *J. Chromatog.*, **61**, 101 (1971).

52. Jansen, L., and Samuelson, O., *J. Chromatog.*, **57**, 353 (1971).

53. Bengtsson, L., and Samuelson, O., *Anal. Chim. Acta*, **57**, 93 (1971); *Chromatographia*, **4**, 142 (1971).

54. Martinsson, E., and Samuelson, O., *Chromatographia*, **3**, 405 (1970).

55. Thacker, L. H., *J. Chromatog.*, **73**, 117 (1971).

56. Katz, S., Pitt, Jr., W. W., and Jones, Jr., G., *Clin. Chem.*, **19**, 817 (1973).

57. Hegenauer, J., and Saltman, P., *J. Chromatog.*, **74**, 133 (1972).

58. Gardell, S., *Acta Chem. Scand.*, **7**, 207 (1953).

59. Crumpton, M. J., *Biochem. J.*, **72**, 479 (1959).

60. Brendel, K., Roszel, N. O., Wheat, R. W., and Davidson, E. A., *Anal. Biochem.*, **18**, 147 (1967).

61. Lee, Y. C., Scocca, J. R., and Muir, L., *Anal. Biochem.*, **27**, 559 (1969).

62. Fanger, M. W., and Smyth, D. G., *Anal. Biochem.*, **34**, 494 (1970).

63. Yaguchi, M., and Perry, M. B., *Can. J. Biochem.*, **48**, 386 (1970).

64. Gregory, J. D., and Van Lenten, L., in *Automation in Analytical Chemistry*, Technicon Symposium 1965, Mediad, Inc., New York, 1966, p. 620.

65. Lee, Y. C., in *Methods in Enzymology*, Academic Press, Inc., 1972 (ed. by V. Ginsburg), Vol. XXVIII, Article 6. Editors-in-chief S. P. Colowick and N. O. Kaplan.

66. Tompsett, S. L., *Anal. Chim. Acta*, **21**, 555 (1959).

67. Hatano, H., Sumizu, K., Rokushika, S., and Murakami, F., *Anal. Biochem.*, **35**, 377 (1970).

68. Morris, D. R., Koffron, K. L., and Okstein, Jr., *Anal. Biochem.*, **30**, 449 (1969).

69. Wall, R. A., *J. Chromatog.*, **60**, 195 (1971).

70. Lepri, L., Desideri, P. G., Coas, V., and Cozzi, D., *J. Chromatog.*, **49**, 239 (1970).

71. Christofferson, K., *Svensk. Papperstidning*, **70**, 540 (1967).

72. Tompkins, E. R., Khym, J. X., and Cohn, W. E., *J. Am. Chem. Soc.*, **69**, 2769 (1947).

73. Cohn, W. E., *Science*, **109**, 377 (1949); *J. Am. Chem. Soc.*, **72**, 1471 (1950).

74. Saukkonen, J. J., in *Chromatographic Reviews*, Elsevier Publishing Co., Amsterdam, 1964 (ed. by M. Lederer), Vol. 6, p. 53.

75. Hurlbert, R. B., Schmitz, H., Brumm, A. F., and Potter, V. R., *J. Biol. Chem.*, **209**, 23 (1954).

76. Crampton, C. F., Frankel, F. R., Benson, A. M., and Wade, A., *Anal. Biochem.*, **1**, 249 (1960).

77. Staehelin, M., Peterson, E. A., and Sober, H. A., *Arch. Biochem. Biophys.*, **85**, 289 (1959).

78. Tener, G. M., Khorana, H. G., Markham, R., and Pol, E. H., *J. Am. Chem. Soc.*, **80**, 6223 (1958).

79. Tomlinson, R. V., and Tenner, G. M. *Biochemistry*, **2**, 697 (1963); *J. Am. Chem. Soc.*, **84**, 2644 (1962).

80. Staehelin, M., in *Progress in Nucleic Acid Research*, Academic Press, Inc., New York, 1963 (ed. by J. N. Davidson and W. E. Cohn), Vol. 2, p. 169.

81. Anderson, N. G., *Anal. Biochem.*, **4**, 269 (1962).

82. Uziel, M., Koh, C. K., and Cohn, W. E., *Anal. Biochem.*, **25**, 77 (1968).

83. Scott, C. D., Jolley, R. L., Pitt, W. W., and Johnson, W. F., *Am. J. Clin. Pathol.*, **53**, 701 (1970); Scott, C. D., Attrill, J. E., and Anderson, N. G., *Proc. Soc. Exp. Biol. Med.*, **125**, 181 (1967); Scott, C. D., Chilcote, D. D., Katz, S., and Pitt, Jr., W. Wilson, *J. Chromatog. Sci.*, **11**, 96 (1973).

84. Kirkland, J. J., *J. Chromatog. Sci.*, **8**, 72 (1970).

85. Kirkland, J. J., *Anal. Chem.*, **43**, 37A (1971).

86. Singhal, R. P., and Cohn, W. E., *Anal. Biochem.*, **45**, 585 (1972).

87. Horvath, C. G., and Lipsky, S. R., *Anal. Chem.*, **41**, 1227 (1969).

88. Murakami, F., Rokushika, S., and Hatano, H., *J. Chromatog.*, **53**, 584 (1970).

89. Cohn, W. E., in *The Nucleic Acids*, Academic Press, Inc., New York, 1955 (ed. by E. Chargaff and J. N. Davidson), Vol. 1, Chap. 6; in *Chromatography*, Reinhold Publishing Corporation, New York, 1967 (ed. by E. Heftmann), 2nd ed., Chap. 22.

90. Blattner, F. R., and Erickson, H. P., *Anal. Biochem.*, **18**, 220 (1967).

91. Cohn, W. E., and Carter, C. E., *J. Am. Chem. Soc.*, **72**, 4273 (1950).

92. Shmukler, H. W., *J. Chromatog. Sci.*, **8**, 653 (1970).

93. Warner, A. H., and Finamore, F. J., *J. Biol. Chem.*, **242**, 1933 (1967).

94. Lapidot, Y., and Barzilay, I., *J. Chromatog.*, **71**, 275 (1972).

95. Horvath, C. G., Preiss, B. A., and Lipsky, S. R., *Anal. Chem.*, **39**, 1422 (1967).

96. Kennedy, W. P., and Lee, J. C., *J. Chromatog.*, **52**, 203 (1970).

97. Brown, P. R., *J. Chromatog.*, **52**, 257 (1970).

98. Bollum, F. J., Groeniger, E., and Yoneda, M., *Proc. Nat. Acad. Sci. USA*, **51**, 853 (1964).

99. Bartos, E. M., Rushizky, G. W., and Sober, H. A., *Biochemistry*, **2**, 1179 (1963).

100. Kothari, R. M., and Taylor, M. W., *J. Chromatog.*, **73**, 449–462 (1972).

101. Kothari, R. M., *Chromatog. Rev.*, **12**, 127 (1970).

102. Mrochek, J. E., Butts, W. C., Rainey, Jr., W. T., and Burtis, C. A., *Clin. Chem.*, **17**, 72 (1971).

103. Stevenson, R. L., and Burtis, C. A., *J. Chromatog.*, **61**, 253 (1971).

104. Bonnelycke, B. E., Dus, K., and Miller, S. L., *Anal. Biochem.*, **27**, 262 (1969).

105. Johnson, W. C., Quill, L. L., and Daniels, F., *Chem. Eng. News*, **25**, 2494 (1947).

106. Rieman III, W., and Walton, H. F., *Ion Exchangers in Analytical Chemistry*, Pergamon Press, New York, 1970, Chap. 8., pp. 149–151.

107a. Minczewski, J., and Dybczynski, R., *J. Chromatog.*, **7**, 98 (1962).

107b. Dybczynski, R., *J. Chromatog.*, **14**, 79 (1964).

107c. Wodkiewicz, L., and Dybczynski, R., *J. Chromatog.*, **32**, 394 (1968).

107d. Wodkiewicz, L., and Dybczynski, R., *J. Chromatog.*, **68**, 131 (1972).

108. Strelow, F. W. E., and Weinert, C. H. S. W., *Talanta*, **17**, 1 (1970).

109. Kraus, K. A., and Nelson, F., *Proc. 1st Intern. Conf. on Peaceful Uses of Atomic Energy*, **7**, 113 (1956); Kraus, K. A., and Nelson, F., *Symposium on Ion Exchange and Chromatography in Analytical Chemistry*, Am. Soc. Testing Mater., Spec. Publ. No. 195 (June 1956).

110. Nelson, F., Murase, T., and Kraus, K. A., *J. Chromatog.*, **13**, 503 (1964).

111. Fritz, J. S., and Rettig, T. A., *Anal. Chem.*, **34**, 1562 (1962).

112. Strelow, F. W. E., Rethemeyer, R., and Bothma, C. J. C., *Anal. Chem.*, **37**, 107 (1965).

113. Faris, J. P., and Warton, J. W., *Anal. Chem.*, **34**, 1077 (1962).

114. Qureshi, M., and Husain, W., *Talanta*, **18**, 399 (1971); Qureshi, M., Husain, W., and Isralli, A. H., *Talanta*, **15**, 789 (1968).

115. Bhatnagar, R. P., Trivedi, R. G., and Bala, Y., *Talanta*, **17**, 249 (1970).

116. Fuller, M. J., *Chromatog. Rev.*, **14**, 45 (1971).

117. Maeck, W. J., Kussy, M. E., and Rein, J. E., *Anal. Chem.*, **35**, 2086 (1963).

118. Nelson, F., Michelson, D. C., Phillips, H. O., and Kraus, K. A., *J. Chromatog.*, **20**, 107 (1965).

119. Dybczynski, R., *J. Chromatog.*, **71**, 507 (1972).

120. Kraus, K. A., Phillips, H. O., Carlson, T. A., and Johnson, J. S., *Proc. 2nd Intern. Conf. on Peaceful Uses of Atomic Energy*, **28**, 3 (1958).

121. Coetzee, C. J., and Rohwer, E. F. C. H., *Anal. Chim. Acta*, **44**, 293 (1969).

122. Donaldson, J. D., and Fuller, M. J., *J. Inorg. Nucl. Chem.*, **32**, 1703 (1970).

123. van Raaphorst, J. G., and Haremaker, H. H., *Talanta*, **17**, 345 (1970).

124. Pollard, F. H., Nickless, G., and Spincer, D., *J. Chromatog.*, **13**, 224 (1964).

125. Araki, S., Suzuki, S., and Yamada, M., *Talanta*, **19**, 577 (1972).

126. Milton, G. M., and Grummitt, W. E., *Can. J. Chem.*, **35**, 541 (1957).

127. Nelson, F., and Kraus, K. A., *J. Am. Chem. Soc.*, **77**, 801 (1955).

128. Morie, G. P., and Sweet, T. R., *J. Chromatog.*, **16**, 201 (1964).

129. Bishay, T. Z., *Anal. Chem.*, **44**, 1087 (1972).

130. Kuroda, R., Kondo, T., and Oguma, K., *Talanta*, **19**, 1043 (1972).

131. Strelow, F. W. E., Rethemeyer, R., and Bothma, C. J. C., *Talanta*, **19**, 1019 (1972).

132. Muzzarelli, R. A. A., in *Advances in Chromatography*, Marcel Dekker, Inc., New York, 1968 (ed. by J. C. Giddings and R. A. Keller), Vol. 5, p. 127.

133. DeGeiso, R. C., Rieman III, W., and Lindenbaum, S., *Anal. Chem.*, **26**, 1840 (1954).

134. Tustanowski, S., *J. Chromatog.*, **31**, 268 (1967).

135. Beukenkamp, J., Rieman III, W., Lindenbaum, S., *Anal. Chem.*, **26**, 509 (1954).

136. Grande, J. A., and Beukenkamp, J., *Anal. Chem.*, **28**, 1497 (1956).

137. Lungren, D. P., and Loeb, N. P., *Anal. Chem.*, **33**, 366 (1961).

138. Kura, G., and Ohashi, S., *J. Chromatog.*, **56**, 111 (1971).

139. Pollard, F. H., Rogers, D. E., Rothwell, M. T., and Nickless, G., *J. Chromatog.*, **9**, 227 (1962).

140. Pollard, F. H., Nickless, G., Rogers, D. E., and Rothwell, M. T., *J. Chromatog.*, **17**, 157 (1965).

141. Skloss, J. L., Hudson, J. A., and Cummishey, C. J., *Anal. Chem.*, **37**, 1240 (1965).

142. Pollard, F. H., Nickless, G., and Glover, R. B., *J. Chromatog.*, **15**, 533 (1964).

Utilization of Ion-Exchange
Resins for Partition, Salting-out, and
Ion-Exclusion Chromatography
EIGHT

A. PRELIMINARY CONSIDERATIONS

Depending upon the nature of a particular solute, there are different compartments available to it as the dissolved solute passes through a high capacity ion-exchange resin. For instance, a large uncharged molecule (e.g., a water-soluble polyalcohol) may simply be too large to penetrate the hydrocarbon network of the exchanger so it passes through the resin bed by flowing in and out of the void spaces in the column, that is, around the exchange particles. A smaller uncharged molecule able to move freely in and out of the exchanger matrix will take longer to appear in the effluent of this same column. As it passes down the column it will have to occupy the volume contained within the exchanger particles as well as in the void spaces of the column; thus the smaller uncharged molecules will be eluted after the larger molecules. For electrolytes the situation is different. Even though small enough to penetrate the interstices of the exchanger matrix, the co-ions of an electrolyte, when present in dilute concentrations, are excluded from the liquid phase inside an exchanger. Thus if a dilute NaCl solution is passed through the Na-form of a cation exchanger or the Cl-form of an anion exchanger, the co-ions in each case (Cl$^-$ in the former case, Na$^+$ in the latter) are excluded. Even though the counter-ions may exchange (Na$^+$ for Na$^+$ in the case of the cation exchanger and Cl$^-$ for Cl$^-$ for the anion exchanger) no net ion exchange occurs, the net effect is that total electrolyte is excluded. This exclusion can be explained on the basis of a Donnan membrane potential (see Chapter 1, and Appendix A) that develops when dilute solutions of electrolytes are brought into contact with a high capacity ion-exchange resin. From the foregoing, for all practical purposes, total electrolyte (regardless of size) behaves as if it is excluded from the interior of an exchanger, while uncharged molecules, provided they are small enough, may freely diffuse in and out of the exchanger matrix. The difference in the ability of an exchanger to exclude electrolytes yet allow nonionized solutes to penetrate its interior is called *ion exclusion*. Perhaps "electrolyte exclusion" would be a more appropriate term.

The technique was first employed for the separation of low molecular-weight ionic and nonionic substances (see Fig. 8-1). Such samples introduced to the resin bed were simply rinsed through the column with water. Since the nonionic compounds have to occupy the total free water contained in the column while the electrolyte is excluded from the interior of the exchanger, the two substances will have different migration rates as they pass through the column. Provided that the sample volume is less than the volume of water inside the resin particles, complete separation occurs (see Fig. 8-1), the ionic substances appearing in the eluate ahead of the uncharged substances.

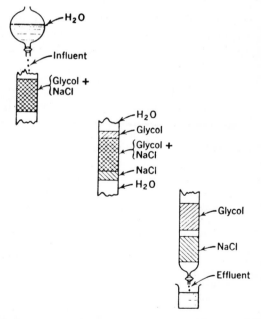

Fig. 8-1. Ion-exclusion separation of ionic from nonionic solutes. (Reproduced from R. M. Wheaton and A. H. Seanster in *Kirk-Othmer Encyclopedia of Chemical Technology*, Interscience Publishers, Inc., New York, 1966, p. 892 by permission of the authors and Interscience Publishers, Inc.)

Theoretically, the electrolyte should appear in the effluent as soon as a volume of water corresponding to the void volume (V_0) has been rinsed through the column. Likewise the nonelectrolyte (assumed small enough to diffuse freely in and out of the resin particle) should appear in the effluent when a volume of water corresponding to V_0 plus V_i (the volume of water inside the exchanger) has been rinsed through the column. If the sample volume is less than V_i, a fraction of eluate containing water only will appear between the separated bands of electrolyte and nonelectrolyte (see Fig. 8-1). In other words, excluding any solute-matrix interactions, the electrolyte should have a volume partition coefficient (K_D) equal to zero and uncharged substances should have the value of unity. In practice these ideal conditions very seldom exist and the elution volume V_e of a solute may be obtained from the fundamental equation of partition chromatography as given by

$$V_e = V_0 + K_D V_i \qquad (8\text{-}1)$$

Almost all the common strong electrolytes are excluded (i.e., have a $K_D = 0$) from ion-exchange resins, but low molecular-weight water-soluble nonelectrolytes have K_D values that range all the way from zero to unity and beyond. The degree of spread for partition coefficients is typical of those shown in Table 8-1. Values between zero and unity mean either partial permeability of solutes into the exchanger particle or that sorption occurs, or both. When the values are higher than unity, sorption is indicated. Also there

TABLE 8-1

Partition Coefficients (K_D) for Some Nonelectrolytes

Compound	Cation exchanger[a]		Anion exchanger[a]	
	H-*form*	Na-*form*	Cl-*form*	SO₄-*form*
Glucose	0.22			
Sucrose	0.24			
Glycerol	0.49	0.56	1.12	
Formaldehyde	0.59		1.06	1.02
Methanol	0.61		0.61	
Ethylene glycol	0.67	0.63		
Acetic acid	0.71			
Acetone	1.20		1.08	0.66
Phenol	3.08		17.7	

[a]Resins were Dowex exchangers, 8% nominal crosslinkage.
Source: Data from R. M. Wheaton and W. C. Bauman, *Ann. N. Y. Acad. Sci.*, 57 (1953). Reproduced, in part, by permission of the authors and the New York Academy of Sciences.

is no sharp distinction between electrolytes and nonelectrolytes (e.g., trichloroacetic acid is separated from acetic acid by ion exclusion). It appears that from the number and kinds of solutes examined, many factors may contribute to the separation of both nonelectrolytes and ionic solutes on ion-exchange resins even exclusive of the mechanism of "true" ion exchange in the case of the latter substances. Some of these are (1) simple partition of a solute between the aqueous and resin phases; (2) the size and shape of solute molecules; (3) aromatic and nonpolar sorption of solute to the hydrocarbon matrix of the exchanger; (4) ionic repulsive forces acting between solutes and the exchange sites of the exchanger; (5) "salting-out" effects when salts or buffer solutions replace water as the eluent in ion-exchange processes whereby the nonelectrolyte increases its sorption by the resin phase; (6) solubility changes when organic solvents are added to eluents (e.g., ethanol-water mixtures) whereby polar nonelectrolytes (e.g., sugars) are relatively more attracted for the more water-rich resin phases; (7) the capacity, crosslinkage, type, functional group, and counter-ion of the ion-exchange resin. Thus only in a theoretical sense is ion exclusion operative exclusive of other chromatographic mechanisms such as ion exchange, partition, adsorption, ion repulsion, and sieving.

Advantage has been taken of these operative factors to achieve some useful and interesting separations. Some widely different examples of separations by ion exclusion and/or related mechanisms are given in the remainder of this chapter. Other examples are given elsewhere by Berg (1), Rieman and Walton (2), Wheaton and Bauman (3), and Sargent and Rieman (4).

B. PARTITION CHROMATOGRAPHY OF CARBOHYDRATES

Separation of sugars by partition chromatography on ion-exchange resins was first described by Samuelson and Sjöström (5) and has been refined by Samuelson and his collaborators in further investigations (6). These workers partitioned sugars between the resin and external solution using aqueous ethanol as the solvent at temperatures around 80°C. An important factor in these partition studies was the observation that the solvent components are distributed unevenly between the resin and the external phase. The relative amount of water is higher in the resin phase. This explains why the polar sugar molecules are attracted preferentially into the exchanger phase. The volume distribution coefficients (D_v)* decrease with decreasing ethanol concentration and with increasing temperature. Another observation is that,

---

*Samuelson and co-workers express their data in terms of D_v as defined in Chapter 2. This is related to the partition coefficient K_D (see Eq. 8-1) by the equation $D_v = K_D \cdot V_i/X$, where V_i is the volume of liquid contained within the exchanger particles and X is the total bed volume of the settled column.

TABLE 8-2

Volume Distribution Coefficients (D_v)[a] of Some Monosaccharides and Anhydrosugars at Various Temperatures and Ethanol Concentrations

Sugars:	Porous resin[b] SO_4^{-2}; 75°	Low-capacity resin[b] SO_4^{-2}; 75°		Dowex 1-X8 SO_4^{-2}; 90°	Dowex 50W-X8 Li^+; 75°	Amberlite IR-120 Li^+	
Ethanol concentrations	88%	86%	90%	86%	92.4%	92.4%; 75°	92.4%; 100°
Erythrose	3.08				1.9		
Threose	3.84				1.4		
Ribose	6.55	4.06	5.80	4.98		4.0	3.1
Lyxose	8.86	5.61	8.70	6.57		2.9	2.4
Arabinose	10.1	6.26	9.79	7.56		3.8	3.0
Xylose	12.5	7.38	12.1	9.19	2.7	3.0	2.4
Fructose	13.5	8.04	13.8	10.3		5.6	4.3
Tagatose		8.04	13.8	9.99		4.5	3.7
Sorbose		8.93	15.6	11.0		4.6	3.7
Mannose	16.4	9.37	16.9	11.8		5.3	4.4
Galactose	23.4	13.0	24.4	16.1	5.7	6.8	5.3
Glucose	28.1	14.9	29.5	19.5	4.8	5.4	4.4
Allose	16.6				6.0		
Altrose	16.6				4.2		
Gulose	20.4				4.9		

Talose	10.7					5.8
Digitoxose						
"2-Deoxyribose"	0.74	0.85	0.72	0.8	0.6	
"2-Deoxygalactose"	1.25	1.57	1.33	1.5	1.3	
"2-Deoxyglucose"		3.26	2.77	2.9	2.3	
	2.80	4.11	3.54	2.1	2.7	
Rhamnose	4.75	2.80	4.11	3.54	1.4	1.2
Fucose	3.25	4.96	4.19	2.4	1.8	
6-Deoxyglucose	7.31					
Levoglucosan (pyranose)	3.84					0.9
Levoglucosan (furanose)	6.29					2.3
						2.4

[a] $D_v = (\bar{v}/X) - \epsilon$ where $\bar{v}$ is the retention volume, X the volume of the resin bed and ϵ the relative interstitial volume.
[b] The porous anion-exchange resins were from various trial batches supplied by Technicon. Source: Reproduced from O. Samuelson in Methods in Carbohydrate Chemistry, Academic Press, Inc., 1972 (ed. by R. L. Whistler and J. N. BeMiller) Vol. VI, Article 9, by permission of the author and Academic Press, Inc.

with few exceptions, D_v increases with the number of hydroxyl groups in a sugar molecule. Also the introduction of nonpolar groups such as methyl groups, as would be expected, results in decreased values for D_v.

A summary of some of this work is given in Table 8-2. If the ratio between D_v's for any pair of compounds is 1.1 or greater, a quantitative separation can be obtained on an analytical column with a length of about 100 cm; as the ratio of D_v's becomes greater, shorter columns may be utilized. This system has been automated by Samuelson and his co-workers and in addition to monosaccharides, higher saccharides, alditols, glycosides, and methylated sugars are separated by partition chromatography on ion-exchange resins. Some modifications of the Samuelson system for the partitioning of carbohydrates on resin columns from ethanol-water mixtures have been made in other laboratories.

Hobbs and Lawrence (7) have introduced the use of organic bases as counter-ions for the separation of carbohydrates on strongly acidic cation-exchange resins with 85% (w/w) ethanol in water as the mobile phase. The best separation of mono- and disaccharides was achieved on the trimethyl-ammonium form of a cation exchanger. A typical separation is demonstrated in Fig. 8-2. As opposed to the automated colorimetry of most carbohydrate analyzers, Hobbs and Lawrence (7) achieved quantitation of the carbohydrate peaks with a moving-wire flame-ionization detector (8).

A very sensitive sugar autoanalyzer that will detect 10^{-10} moles of carbohydrate and complete an analysis in 3 to 4 hours has been developed by Mopper and Degans (9). Their column is filled with the SO_4-form of Technicon type S resin (10% crosslinked styrene-divinylbenzene beads, 20 μ in

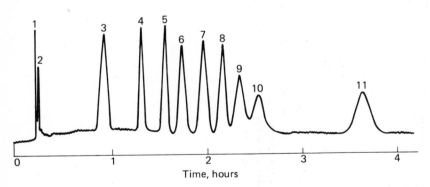

Fig. 8-2. Separation of mono- and disaccharides. Column, 0.4×100 cm, Aminex A-6, trimethylammonium form; eluent, 85% (w/w) ethanol in water; temperature, 65°C; flow rate, 0.266 ml min^{-1}. Each peak corresponds to approximately 150 μg carbohydrate. 1 = methyl palmitate; 2 = tetramethylglucose; 3 = rhamnose; 4 = ribose; 5 = xylose; 6 = arabinose; 7 = mannose; 8 = glucose; 9 = galactose; 10 = maltose; 11 = lactose. (Reproduced from J. S. Hobbs and J. G. Lawrence, *J. Chromatog.*, **72**, 314 (1972), by permission of the authors and Elsevier Publishing Company.)

diameter) and is operated at 400 psi with 89% ethanol as the solvent at a flow rate of about 0.5 ml per min. The column heating jacket is maintained at 75°C with a constant-temperature circulator. The eluate was monitored continuously for sugar content by supplying an alkaline solution of a dye, tetrazolium blue, to the effluent stream. This colorimetric method of detecting sugars in an automatic system avoids the difficulties involved in manipulating corrosive mixtures (e.g., orcinol-sulfuric acid, anthrone-sulfuric acid, and phenol-sulfuric acid) that are usually employed in autoanalyzers to detect sugars. Long columns (> 100 cm) yielded the best resolution and narrow diameters (< 0.3 cm) yielded the greatest sensitivity. The general order of elution is: deoxypentoses, pentoses, deoxyhexoses, hexoses, and disaccharides.

Saunders (10) has developed a partition method for the separation of sugars on the K-form of a strong-acid cation exchanger. Water only is used to chromatograph sugars on a column of Dowex 50W (K-form). Sugars (25–150 mg) were applied to the column in about 5 ml of water. The aqueous fractions were collected with an automatic fraction collector and the sugars in the eluate were determined by the phenol-sulfuric acid method. As seen in Fig. 8-3, fractionation occurs according to molecular size. Thus far the procedure has been only applied to relatively simple mixtures of carbohydrates.

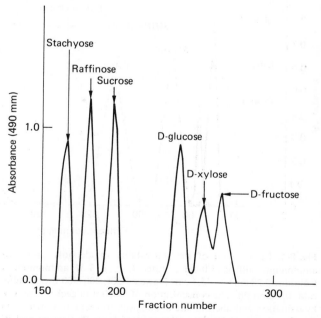

Fig. 8-3. Separation of sugars on cation exchanger. Conditions: column, 4.5 × 120 cm of Dowex 50W, K-form; elution with water at 0.8 ml/min, 25°C. (Reproduced from R. M. Saunders, *Carbohydr. Res.*, **7**, 76 (1968), by permission of the author and Elsevier Publishing Company.)

C. SALTING-OUT CHROMATOGRAPHY

The separation of nonionic substances from each other on resinous exchangers may be facilitated if salt solution is used as the eluting agent instead of water. The presence of salt increases the sorption of the nonelectrolyte into the resin phase. The degree of sorption is different for different solutes, the hydrophobic nature of the nonionic solute probably being the determining factor, since an added electrolyte "salts out" the nonelectrolyte component from the hydrophillic environment of the aqueous phase to the partial hydrophobic environment of the resin phase. This type of separation is called *salting-out chromatography* (2). The salt chosen for the eluent should: (1) not interfere with the determination of the solute; (2) be very soluble in water; (3) exert a large salting-out effect. Ammonium sulfate fulfills these requirements and has been used extensively in salting-out chromatography (2, 4).

Typical of the effects noted when salt replaces water as the eluent is demonstrated in Fig. 8-4. Sargent and Rieman (4) showed that the simple aliphatic alcohols could be separated on the sulfate form of Dowex 1-X8 in

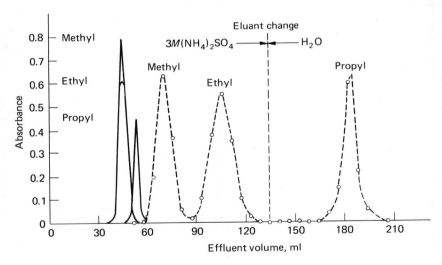

Fig. 8-4. Comparison of the separability of alcohols with water and $3.0\,M$ ammonium sulfate. Eluents: water (⎯⎯⎯), $3\,M$ ammonium sulfate (_ _ _ _). Column: $2.28\,cm^2 \times 25.7\,cm$, Dowex 1-X4, 200–300 mesh, sulfate form. Flow rate: 0.5 cm/min. The concentration of alcohol in each fraction was determined by oxidation with dichromate in 50% sulfuric acid and then by measurement of the absorbance of Cr (III). (Reproduced from R. Sargent and W. Rieman III, *J. Org. Chem.*, **21**, 594 (1956), by permission of the authors and American Chemical Society.)

the presence of 3 M ammonium sulfate. The solid curves represent the elution position of the alcohols when water is used as the eluent, and the dotted curves represent the elution position when water is replaced by ammonium sulfate solution. Propyl alcohol sorbs so strongly to the exchanger in the presence of salt (eluent position about 475 ml), that water is used to remove it once methyl and ethyl alcohol have been separated.

In addition to the separation of alcohols, salting-out chromatography has been applied for the fractionation of other compounds such as aldehydes, ketones, ethers, esters, or aromatic hydrocarbons (2). This technique is particularly effective for the separation of isomers. Funasaka *et al.* (11) separated the hydroxybenzoic acid isomers by salting-out chromatography. These workers utilized a weakly acidic carboxylic acid exchanger (Amberlite CG-50, Na-form) as the stationary phase and acidic sodium chloride as the mobile phase. The best separation was obtained when a mixture of the *o*- and *m*- or the *o*- and *p*-isomers was eluted at 40°C through a 1 × 15 cm column with 2 M sodium chloride of pH 1.8. At flow rates of 0.2 to 0.4 ml per min, quantitative separation and recovery of microgram quantities of isomeric mixtures was readily achieved. The concentration of the acids in the eluate was determined by UV spectrophotometry. Gradient elution salting-out chromatography was applied to the analysis of alkylsulfate-soap mixtures by Fudano and Konishi (12). Soap samples were solubilized in the initial eluent of 30% isopropanol-0.5 sodium chloride solution and applied to a

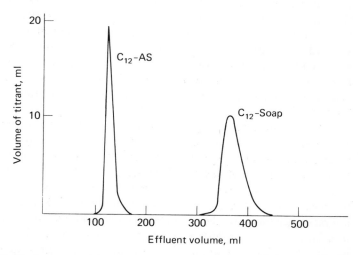

Fig. 8-5. Elution curve of C_{12}-alkyl sulfate and C_{12}-soap. Conditions: Column, 2.5 × 23.5 cm of Amberlite CG-50, Na-form; elution with NaCl-isopropanol solutions as explained in text at 0.3 ml/min, 40°C; sample size, 20 mg each component. (Reproduced from S. Fudano and K. Konishi, *J. Chromatog.*, **71**, 93 (1972), by permission of the authors and Elsevier Publishing Company.)

weakly acidic cation-exchange column. Simultaneously with the start of elution, 30 % isopropanol was fed into the initial eluting solution. Consequently the concentration of sodium chloride in the eluent decreased gradually as the elution progressed. As shown in Fig. 8-5, this allowed the complete separation of C_{12}-alkylsulfate from C_{12}-soap. Quantitative analyses of eluate fractions were carried out by a two-phase titration method with bromocresol green as the indicator.

D. SEPARATION OF IONIC COMPOUNDS BY ION EXCLUSION AND RELATED METHODS

In the main, ion exclusion has been applied only for group separations of ionic from nonionic substances. As demonstrated earlier in this chapter, several mechanisms have been exploited that have led to chromatographic procedures for the separation of nonionized solutes on the resinous exchangers. However, within the realm of the concept of ion exclusion the same effort has not been applied to the chromatographic separation of ionized compounds on the resinous exchangers. That is, "nonionexchange" mechanisms for the separation of ionized solutes on the resinous polymers have not been studied as extensively. Three different examples of this latter type of chromatography are given here.

Simple examples of this are shown by the ion-exclusion separation of stronger electrolytes from weaker ones. In these cases, the latter substances behave as though they were the nonelectrolyte component of an ion-exclusion separation. Wheaton and Bauman (3) give data to show that even certain weak acids of different acid strength can be separated by ion exclusion. In most cases, however, these are simple binary mixtures (e.g., chloroacetic and dichloroacetic acids) and overall, in this area, the applications are very limited.

A more striking example of separating ionic materials through nonionexchange mechanisms on resinous exchangers is given by Seki's work (13), whereby he separated phenol and its derivatives on the Na-form of Amberlite IR 112. He used a carbonate buffer system at pH 9.7 to obtain the separation of phenol and the three isomeric cresols that are shown in Fig. 8-6. At pH 9.7 these phenol derivatives exist as partially ionized negative species and, at first glance, one would expect that these compounds would not be attracted to a cation exchanger. However, aromatic compounds exhibit a strong affinity for the hydrocarbon matrix material of the resin and sorption takes place by hydrophobic interactions due to van der Waals and nonpolar forces. Since the ionic functional groups of the cation exchanger bear negative charges as do the phenolic compounds at pH 9.7, desorption must occur through negative-ion repulsive forces. Differences in the balance between van der Waals and these repulsive forces, among each individual phenol

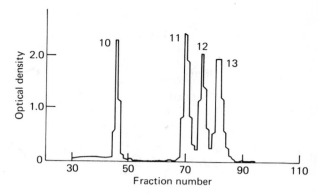

Fig. 8-6. Elution curve of phenol and the three isomers of cresol. Column: 0.8 × 150 cm, Amberlite 112, Na-form. Elution with sodium carbonate—EDTA buffer of pH 9.7 (rate not given). The compounds in the order of their elution from the column are: phenol (10), *m*-cresol (11), *o*-cresol (12) and *p*-cresol (13). (Reproduced from T. Seki, *J. Chromatog.*, **4**, 6 (1960), by permission of the author and Elsevier Publishing Company.)

derivative, allow the chromatographic separation to take place. As would be expected, the compounds are eluted in the order of their increasing pK values except for the interchange of *o*- and *p*-cresol. These latter two phenolic derivatives have pK values that are barely distinguishable, the value for the *p*-isomer being 10.26 and that of the *o*-isomer 10.29. In this case, the structural difference presumably has a greater influence in establishing elution position due to hydrophobic binding than does the difference in the degree of ionization of the two isomers.

This same type of chromatography has been used by Lange and Hempel (14) to separate a wide range of aromatic compounds (acids, aldehydes, alcohols). In this case, the aromatic substances were resolved on a strong-acid cation exchanger with sodium citrate-boric acid buffers of pH's between 3.28 and 4.53. Figure 8-7 shows the chromatogram for the separation of ten phenolic acids by this method. Again, as indicated previously, this type of chromatography is not a cation-exchange process but is a form of sorption chromatography that occurs due to the attraction of the aromatic acids to the styrene-divinylbenzene matrix of the cation exchanger. Desorption is apparently influenced by the negativity of the carboxylic groups of these compounds, since the aromatic acids are eluted in the order of their increasing pK values.

The most elegant demonstration of chromatographic separation of ionic substances on ion-exchange resins by mechanisms other than ion exchange is given by the work of Singhal (15) and Singhal and Cohn (16, 17). With very few exceptions (e.g., the behavior of a few simple weak electrolytes) heretofore, only nonionic solutes have exhibited the property of partial exclusion

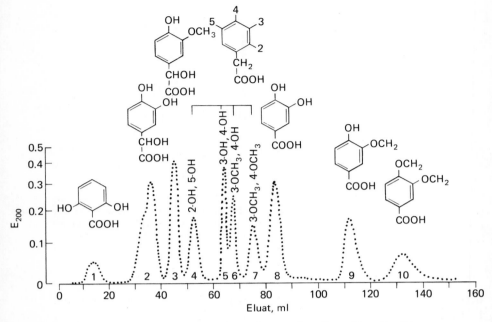

Fig. 8-7. Chromatographic separation of ten phenolic acids. Conditions: Column, 0.9 × 60 cm of cation-exchange resin PA-28 (a sulfonated styrene-divinylbenzene polymer obtained in Germany), 16 ± 6 μ beads; eluent, 0–15 min Na-citrate buffer, pH 3.28, 15–150 min Na-citrate-boric acid buffer, pH 4.53; flow rate, 50 ml/h; pressure 15–20 atm; column temperature, 55°C. 1 = 2, 6-dihydroxybenzoic acid (200 μg); 2 = 3,4-dihydroxymandelic acid (600 μg); 3 = 3-methoxy-4-hydroxymandelic acid (400 μg); 4 = homogentisic acid (200 μg); 5 = homoprotocatechuic acid (200 μg); 6 = homovanillic acid (200 μg); 7 = homoveratric acid (200 μg); 8 = protocatechuic acid (200 μg); 9 = vanillic acid (200 μg); 10 = veratric acid (200 μg). (Reproduced from H. W. Lange and K. Hempel, *J. Chromatog.*, **59**, 53 (1971), by permission of the authors and Elsevier Publishing Company.)

from a resin exchanger; that is, have appeared in less than one liquid column volume of eluent. Singhal and Cohn have demonstrated that low molecular-weight nucleic acid components under certain conditions will also appear within elution volumes less than the liquid content of a resin column. These workers discovered that differential partial ion exclusion among nucleic acid components took place on the resin exchangers when the compounds of interest were eluted at the "wrong" pH, i.e., by eluting anionic species from a cation exchanger at an alkaline pH and by eluting cationic species from anion exchangers at an acid pH. The separation of a mixture of nucleosides on the NH_4-form of a cation exchanger at alkaline pH is demonstrated in Fig. 8-8. The four components appearing in the effluent in less than one liquid column

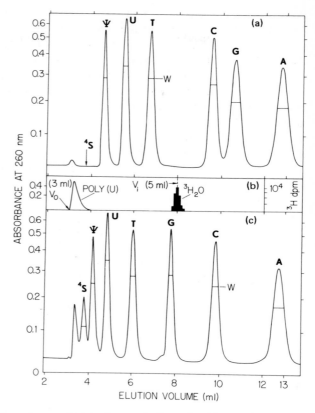

Fig. 8-8. Separation of an artificial mixture of ribonucleosides by chromatography on a sulfonated polystyrene (cation-exchange) column. Column: BioRad Aminex A-6, 0.5 × 50 cm. A total of about 100 moles ($\simeq 0.8\,A_{260\,nm}$) unit) mixture of 4-thiouridine (^{4}S), pseudouridine (ψ), uridine (U), 5-methyluridine or ribothymidine (T), guanosine (G), cytidine (C), and adenosine (A) in 10 μl buffer solution was applied to the column prior to the start of the elution by 0.02 M $(NH_4)_2CO_3$ + NH_4OH at pH 9.3 [Panel (a)] or at pH 9.85 [Panel (c)] at 0.98 ml/cm^2 per min or 0.194 ml/min, and at 50°C. The peak prior to 4-thiouridine in Panel (c) corresponds to sulfonic acid, a contaminant in the sample. W's in the Panels (a) and (c) represent half-widths at maximum peak heights. Panel (b) shows the eluting positions of poly (U) and of ^3H_2O on the same column, marking the void volume (V_0, liquid between the beads) and the total volume (V_i, liquid within the beads), respectively. This column represents a total liquid volume ($V_0 + V_i$) of 8.0 ml for one bed volume (9.8 ml). It is apparent from this figure that the five substances having pK values in the region of 9 are variably repulsed or excluded from the negatively charged resin depending upon the degree to which their negative charges are affected by the pH of the eluting solution. (Reproduced from R. P. Singhal and W. E. Cohn, *Biochim. Biophys. Acta*, **262**, 565 (1972), by permission of the authors and Elsevier Publishing Company.)

volume of eluate (see Fig. 8-8) are eluted in the increasing order of their pK values. It seems apparent that ion exclusion (probably due to ion repulsion) has taken place. A small increase in pH from 9.3 to 9.8 [compare Panels (a) and (c) of Fig. 8-8] yields a net increase in the ionization of some of the nucleosides and they tend to elute sooner than before. The degree of ionization in a substance like guanosine (G), having a pK of 9.2, is increased markedly and it now appears before cytidine (C). The latter nucleoside and adenosine (A) are not ionized at these pH values and their elution position is not affected by changes in pH. Their elution position is established presumably on the basis of simple partition chromatography. Nanomole amounts of 4-thiouridine, pseudouridine, uridine, ribothymidine, guanosine, cytidine, and adenosine are resolved by this method in one hour or less utilizing volatile eluents (0.02 M (NH$_4$)$_2$CO$_3$, pH 9.3 or 9.8). Details as to the influence of the separation parameters on these nucleosides and other nucleic acid components are given by Singhal (15). Cation-exclusion chromatography of nucleic acid components on anion exchangers has also been reported by Singhal and Cohn (17). These studies have added a new dimension to ion-exclusion work, namely that of ion-exclusion chromatography.

Perhaps the most unusual application of "wrong-way" chromatography is seen in a report by Kothari (18). This worker utilized the Al-form of Am-

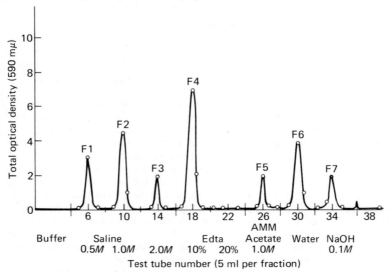

Fig. 8-9. Typical chromatographic elution profile of buffalo liver DNA. Exchanger, 10 g of dry Amberlite 1R-120 (Na-form) converted to Al-form in a column (dimensions not given). Test material sorbed from glycine-sodium hydroxide buffer (pH 8.6, 0.054 M). Elution as shown at 10–15 ml/hr. (Reproduced from R. M. Kothari, *J. Chromatog.*, **52**, 119 (1970), by permission of the author and Elsevier Publishing Company.)

berlite IR-120, a polystyrene sulfonate type of cation exchanger, for the fractionation of deoxyribonucleic acid (DNA), a large molecular-weight polyphosphate polymer found in living organisms. As seen in Fig. 8-9, the Al-form exchanger resolves the DNA into seven distinct fractions and each of these was shown to differ in base composition. The separation is quantitative and reproducible. As shown in Fig. 8-9, each DNA fraction is eluted stepwise with eluents of widely differing composition. Thus, several mechanisms are probably involved in the desorption processes of this single-column method such as complex-ion (Al-DNA) exchange, salting-out, ligand-exchange, and partition chromatography.

REFERENCES

1. Berg, E. W., *Physical and Chemical Methods of Separation*, McGraw-Hill Book Company, Inc., New York, 1963, Chap. 11.

2. Rieman III, W., and Walton, H. F., *Ion Exchangers in Analytical Chemistry*, Pergamon Press, New York, 1970, Chaps. 2 and 9.

3. Wheaton, R. M., and Bauman, W. C., *Ind. Eng. Chem.*, **45**, 228 (1953).

4. Sargent, R., and Rieman III, W., *J. Org. Chem.*, **21**, 594 (1956); *J. Phys. Chem.*, **61**, 354 (1957); *Anal. Chim. Acta*, **17**, 408 (1957); see also Rieman III, W., and Sargent, R., in *Physical Methods in Chemical Analysis*, Academic Press, Inc., New York, 1961 (ed. by W. G. Berl), pp. 189–207.

5. Samuelson, O., and Sjöström, E., *Svensk. Kem. Tidskr.*, **64**, 305 (1952).

6. Samuelson, O., in *Ion Exchange*, Marcel Dekker, Inc., New York, 1969 (ed. by J. A. Marinsky), Vol. 2, Chap. 5; in *Methods in Carbohydrate Chemistry*, Academic Press, Inc., 1972 (ed. by R. L. Whistler and James N. BeMiller) Vol. VI, Article 9.

7. Hobbs, J. S., and Lawrence, J. G., *J. Chromatog.*, **72**, 311 (1972).

8. Hobbs, J. S., and Lawrence, J. G., *J. Sci. Food Agr.*, **23**, 45 (1972).

9. Mopper, K., and Degens, E. T., *Anal. Biochem.*, **45**, 147 (1972).

10. Saunders, R. M., *Carbohyd. Res.*, **7**, 76 (1968).

11. Funasaka, W., Fujimura, K., and Kushida, S., *J. Chromatog.*, **64**, 95 (1972).

12. Fudano, S., and Konishi, K., *J. Chromatog.*, **71**, 93 (1972).

13. Seki, T., *J. Chromatog.*, **4**, 6 (1960).

14. Lange, H. W., and Hempel, K., *J. Chromatog.*, **59**, 53 (1971).

15. Singhal, R. P., *Arch. Biochem. Biophys.*, **152**, 800 (1972).

16. Singhal, R. P., and Cohn, W. E., *Biochim. Biophys. Acta*, **262**, 565 (1972).

17. Singhal, R. P., and Cohn, W. E., *Biochemistry*, **12**, 1532 (1973).

18. Kothari, R. M., *J. Chromatog.*, **52**, 119 (1970).

Ligand-Exchange Chromatography
NINE

A. DESCRIPTION OF LIGAND EXCHANGE

Cation exchangers containing counter-ions such as Cu^{++}, Ni^{++} or Ag^+ show a unique preference for sorbing ammonia, amines, amino acids, or other molecules that can act as ligands (1). Thus even though they are bound to an exchanger, these metals, particularly transition elements, retain their ability to be the central atom of a coordination compound. Furthermore, the ligand associated with the coordinating metal attached to the resin can be replaced by a different ligand, as seen by the example give in the equation

$$(R^-)_2[Cu(NH_3)_4{}^{++}] + 4\,CH_3NH_2 \rightleftharpoons (R^-)_2[Cu(NH_2CH_3)_4{}^{++}] + 4\,NH_3$$

$$(9\text{-}1)$$

where R represents an exchange site of the resin. No ion exchange takes place, the exchanger acts simply as a solid inert support for the complexing metal ion. This process was termed *ligand exchange* by Helfferich (1) who first demonstrated this phenomenon by using a nickel-loaded carboxylic acid exchanger for replacement of amines by elution with ammonia (2). Ligand-exchange reactions have also been studied extensively by Walton and his co-workers (3) who compared several resin types and coordinating metals in studies of ligand exchange between amines or amine derivatives and ammonia. The details and results of these studies are given elsewhere (3).

Although the ordinary strongly acidic and weakly acidic cation exchangers undergo very satisfactory ligand-exchange reactions, the chelating resins seem ideally suited for ligand-exchange work. The chelating polymers have iminodiacetate functional groups attached to a styrene-divinylbenzene matrix. Thus, divalent metal ions, particularly strong complex formers such as nickel or copper, are tightly bound to such exchangers. Consequently, leakage of metal ions from the chelating resins by ordinary ion-exchange reactions with cationic materials (e.g., ammonium ion) in eluting solutions is held to a minimum.

Ligands, for example, amines or amino acids, may be separated from each other by ligand-exchange chromatography. The mechanism for separation is very similar to that of ordinary exchange, namely, different ligands have different affinities for the coordinating metal attached to the exchanger. Hence, their migration rates down a column differ and thus separation occurs. Ligand exchange is a highly selective process and even very similar ligands, under a proper set of conditions, may exhibit differences in the degree of formation of their metal-ligand complexes. Many investigators have taken advantage of these differences to develop ligand-exchange chromatographic procedures that will separate members among a group of closely related substances. Some of this work is described in the remainder of the chapter.

B. SEPARATION BY LIGAND-EXCHANGE CHROMATOGRAPHY

1. Amphetamine Drugs

Based on the previous work (3) of separating aliphatic amines by ligand exchange, Hernandez and Walton (4) applied this technique for the resolution of mixtures of amphetamine bases. The principle amphetamine drugs and related compounds studied were mixtures of

Phenethylamine	$C_6H_5CH_2CH_2NH_2$
Amphetamine	$C_6H_5CH_2 \cdot CHCH_3 \cdot NH_2$
Metamphetamine	$C_6H_5CH_2 \cdot CHCH_3 \cdot NHCH_3$
Norephedrine	$C_6H_5CHOH \cdot CHCH_3 \cdot NH_2$
Ephedrine	$C_6H_5CHOH \cdot CHCH_3 \cdot NHCH_3$

The column material giving the best resolution of these mixtures was a copper-loaded carboxylic acid exchanger. Eluents were aqueous alcohol solutions containing ammonia. A separation, representative of this work, is shown in Fig. 9-1.

2. Purine and Pyrimidine Derivatives

Goldstein (5) demonstrated that nucleic acid components can be separated on the cupric form of a chelating resin. Weakly sorbed components such as uridine, xanthosine, and their derivatives are separated by elution with only

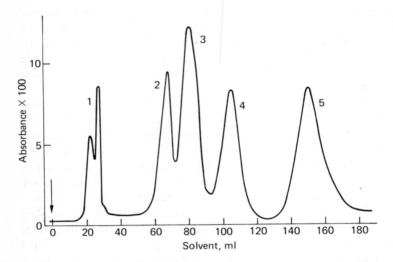

Fig. 9-1. Separation of amphetamine drugs. Conditions: Column, 0.9×60 cm of Bio-Rex 70, copper form; elution with $0.10 \, N$ NH$_4$OH—33% ethanol at 12 ml/hr. Test material: (1) decomposition products; (2) metamphetamine; (3) amphetamine and epherdrine each 1.5 mg/ml; (4) norephedrine; and (5) phenethylamine each 2.5 mg/ml. (Reproduced from C. M. de Hernandez and H. F. Walton, *Anal. Chem.*, **44**, 890 (1972), by permission of the authors and the American Chemical Society.)

water. Stronger sorbed compounds such as some of the common nucleosides of ribonucleic acids, as shown in Fig. 9-2, are effectively separated with $1 \, N$ ammonium hydroxide as the eluent. The purine and pyrimidine bases derived from these four nucleosides are similarly separated by ligand exchange with $2.5 \, N$ ammonium hydroxide as the eluent.

Nucleotides are not sorbed to the copper-loaded ligand column and are washed from the column with water and appear in a sharp band at the breakthrough volume. This observation led to a procedure developed by Burtis and Goldstein (6) for the terminal nucleoside assay of transfer ribonucleic acid (tRNA). Intact tRNA species terminate in an adenosine residue. When hydrolyzed with alkali, the adenosine residue should appear as free nucleoside in a mixture of 2'- and 3'-nucleotides at a concentration of about $\frac{1}{80}$ the molar concentration of the nucleotides. The determination of adenosine in such a mixture is rather difficult. The analysis of these mixtures by ligand exchange offers a convenient method for the terminal assay of tRNA's, since, as shown in Fig. 9-3, adenosine is widely separated from the more concentrated 2'- and 3'-nucleotides. The ligand-exchange method is more sensitive and is not as lengthy as most other analytical procedures for terminal-assay determinations. With a suitably sensitive ultraviolet detector and recorder, terminal nucleoside analysis of tRNA by ligand exchange can

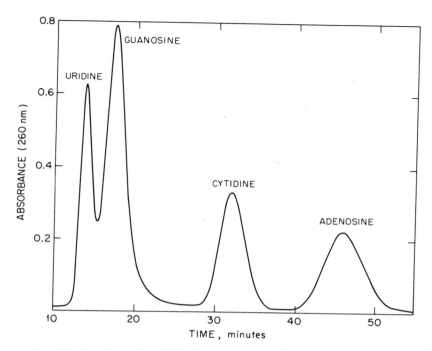

Fig. 9-2. Separation of uridine, guanosine, cytidine, and adenosine. Conditions: Column, 0.9 × 33 cm of Chelex 100, copper form; elution with 1 N NH$_4$OH at 0.77 ml/min. (Reproduced from G. Goldstein, *Anal. Biochem.*, **20**, 477 (1967), by permission of the author and Academic Press, Inc.)

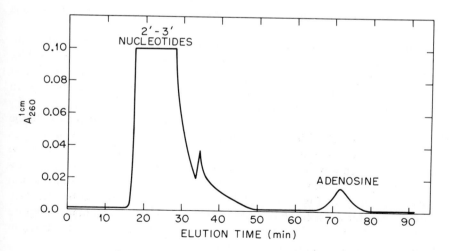

Fig. 9-3. Ligand-exchange chromatogram of alkaline hydrolyzate of tRNA. Conditions: column, 0.9 × 50 cm of Chelex 100, copper form; elution with 1 N NH$_4$OH at 0.77 ml/min. (Reproduced from C. A. Burtis and G. Goldstein, *Anal. Biochem.*, **23**, 502 (1968), by permission of authors and Academic Press, Inc.)

227

be effected on as little as 10 nanomoles of free nucleoside in about one hour with an accuracy of 3–4%.

In similar work, several oxypurines, particularly the methylated derivatives of xanthine and guanosine, were separated by ligand-exchange chromatography on the cupric form of a chelating resin (7). In this case, 1 M ammonium hydroxide served as the eluent for a 1 × 80 cm column operated at a flow rate of 0.75 ml per min.

3. Peptides and Amino Acids

Several reports in the literature refer to the separation of amino acids and peptides by ligand-exchange reactions. The first report to be described here fits the concept of ligand exchange as previously given in Section A of this chapter. In the others, the separations, although referred to as ligand exchange in the original publications, appear to be based on ion exchange rather than ligand exchange. In these latter reports another counter-ligand (such as ammonia) is not present in the buffered eluents used to desorb charged amino acid (or peptide)-metal complexes. During migration of the sample down the metal-loaded column, separation is effected by the differences in the strength of adduct formation of the amino acids (or peptides) with the metallic ion sorbed on the resin, the complexes of lower stability eluting first. These types of analyses are included in this section because they are closely associated with ligand-exchange work and because they are unique in comparison with the well documented conventional ion-exchange techniques (see Chapter 7) usually employed for the separation of amino acids and peptides.

Boisseau and Jouan (8) fractionated mixtures of oligopeptides and amino acids on the Cu-form of the iminodiaceto chelating resin, Chelex X-100. After equilibration of the metal-loaded column with ammonia solution at pH 10.3, acidic and neutral peptides and acidic amino acids are eluted with water. Neutral amino acids and basic peptides are eluted with 1.5 M ammonium hydroxide, and tryptophan and basic amino acids such as histidine and arginine emerge with 6 M ammonia solution. These groups of substances are quantitatively separated as is demonstrated in Fig. 9-4. Separation of the members of each group is possible by rechromatography of the isolated fractions on the same column. This is shown in Fig. 9-5 wherein the water fraction (peaks A and B) of Fig. 9-4 is resolved upon elution of the sorbed fraction with water after the column had been equilibrated with ammoniacal solution of pH 8.3. More strongly sorbed fractions are resolved with ammonium hydroxide solution as the eluting agent.

The separation of peptides from amino acids in urine by a form of ligand exchange was achieved by Buist and O'Brien (9). Urine samples containing 0.03 to 0.07 millimole of amino acids and a similar quantity of peptides at pH 11 were applied to a column (1.3 × 12 cm) of a Chelex resin in

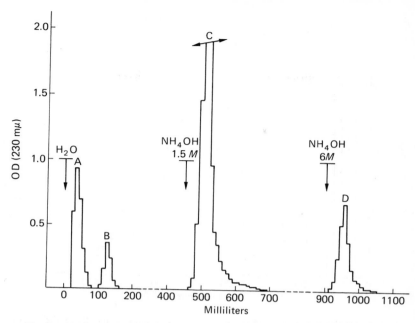

Fig. 9-4. Separation of amino acids and oligopeptides on the copper form of Chelex 100 after equilibration of the column (1.5 × 27 cm) with ammoniacal solution of pH 10.3. Elution, 20 ml/hr with the eluents as shown. Fraction A: aspartic and glutamic acids, and the peptides glutathion, glycyl-alanine, glycyl-tyrosine, and prolyl-glycyl-phenylalanine. Fraction B: glycyl-tryptophan. Fraction C: neutral amino acids (excepting tryptophan), lysine, and the peptides glycyl-histidine, glycyl-lysine, and carnosine. Fraction D: tryptophan, histidine, and arginine. (Reproduced from J. Boisseau and P. Jouan, *J. Chromatog.*, **54**, 231 (1971), by permission of the authors and the Elsevier Publishing Co.)

the copper form. Samples were rinsed on the column with borate buffer at pH 11 until 50 ml of eluate was obtained. No amino acids appeared in the column eluate, while dipeptides and polypeptides were not retained by the resin and were recovered from the effluent along with large quantities of copper, which was subsequently removed by an extraction procedure. The authors state the reason is unclear why the amino acids are retained by the column and the peptides are not, but the investigators suggest that the fact that peptides will strip copper from the resin seems to indicate that peptide bonds have an even stronger affinity for the metal than do the iminodiacetic groups of the resin. It is further suggested that the peptide-copper complex is then configurationally unable to form a further ligand bond between its free amino groups and the copper on the resin.

A commercially available instrument (the Hitachi Perkin-Elmer amino acid analyzer) separates amino acids on the basis of the differences in affinities of the zinc complexes of amino acids towards a sulfonated polystyrene resin.

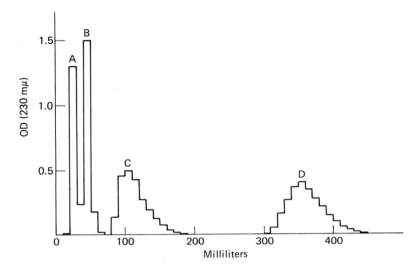

Fig. 9-5. Rechromatography of the water fraction (peaks A and B) of Fig. 9-4 after equilibration of the same Chelex column with ammoniacal solution of pH 8.3. Elution with water. Fraction A: glutathione; Fraction B: glycyl-alanine; Fraction C: glycyl-tyrosine; Fraction D: glycyl-tryptophan. (Reproduced from J. Boisseau and P. Jouan, *J. Chromatog.*, **54**, 231 (1971), by permission of the authors and the Elsevier Publishing Co.)

The acid and neutral amino acids are eluted from a 0.9 × 60 cm column of resin equilibrated at 56°C with buffer of pH 4.08 which contains 55 mM sodium acetate and 0.4 mM zinc acetate. The basic amino acids are eluted from a 0.9 × 22 cm column equilibrated at 60°C with sodium acetate buffer that may be varied with respect to pH and sodium ion concentration, but has a constant zinc acetate concentration of 1 mM. Wagner and Shepherd (10) use this technique and variations of it to separate amino sugars and cysteine derivatives when present in amino acid mixtures. In similar work from this same laboratory (11), γ-aminobutyric acid at very low levels and in the presence of other amino acids was detected in physiological fluids by chromatographic analysis of its zinc complex.

REFERENCES

1. Helfferich, F., *Nature*, **189**, 1001 (1961).

2. Helfferich, F., *J. Am. Chem. Soc.*, **84**, 3237 (1962); **84**, 3242 (1962).

3. Shimomura, K., Dickson, L., and Walton, H. F., *Anal. Chim. Acta*, **37**, 102 (1967); Hill, A. G., Sedgley, R., and Walton, H. F., *Anal. Chim. Acta*, **33**, 84 (1965); Latterell, J. J., and Walton, H. F., *Anal. Chim. Acta*, **32**, 101 (1965).

4. de Hernandez, C. M., and Walton, H. F., *Anal. Chem.*, **44**, 890 (1972).

5. Goldstein, G., *Anal. Biochem.*, **20**, 477 (1967).

6. Burtis, C. A., and Goldstein, G., *Anal. Biochem.*, **23**, 502 (1968).

7. Wolford, J. C., Dean, J. A., and Goldstein, G., *J. Chromatog.*, **62**, 148 (1971).

8. Boisseau, J., and Jouan, P., *J. Chromatog.*, **54**, 231 (1971).

9. Buist, N. R. M., and O'Brien, D., *J. Chromatog.*, **29**, 398 (1967).

10. Wagner, F., and Shepherd, S. L., *Anal. Biochem.*, **41**, 314 (1971).

11. Wagner, F., and Liliedahl, R. L., *J. Chromatog.*, **71**, 567 (1972).

APPENDIX

A. Extension of the Donnan Principle

As explained in Chapter 1, when an exchanger is immersed in a solution of electrolyte, the concentration of free electrolyte, at equilibrium, will be less inside the resin bead than it is in the external solution. Simple examples were used to illustrate this concept; however, even though several diffusible membrane-penetrating electrolytes may be in contact with a resin particle, the Donnan effect still prevails. The ion product concentration of any pair of univalent and oppositely charged ions will have the same value for both sides of a membrane separating a solution that has attained equilibrium and this is irrespective of the number and nature of other ions that may be present. This can be illustrated in the following example. Suppose a second diffusible cation K^+ is added to the system of Fig. 1-6 of Chapter 1. By the Donnan theory

$$[Na^+]_1[Cl^-]_1 = [Na^+]_2[Cl^-]_2 \tag{A-1}$$

and

$$[K^+]_1[Cl^-]_1 = [K^+]_2[Cl^-]_2 \tag{A-2}$$

Division of Eq. A-1 by Eq. A-2 gives the following expression

$$\frac{[Na^+]_1}{[K^+]_1} = \frac{[Na^+]_2}{[K^+]_2} \tag{A-3}$$

Accordingly, an exchange of these cations will take place until their concentration ratio is equal in both the resin phase and in the solution surrounding the resin bead. And, of course, for an anion exchanger, the Donnan effect would govern the distribution of anions between both phases and negative ions would be involved in an expression such as that given for the cations of Eq. A-3. It is on the basis of these considerations and of those given in Chapter 1 that an anion or a cation exchanger exhibits its specificity, respectively, for either anions or cations. Thus, although co-ions (at low concentrations) are prevented from penetrating the resin matrix, no such restriction applies to the counter-ions of an exchanger or to identical ions in the external solution, which are free to diffuse back and forth into either phase and are able to undergo exchange reactions. This, of course, is the manner in which exchanges take place between different monvalent species in equilibrium with an exchanger.

For the case of polyvalent ions, the valence factor has to be considered in setting up the Donnan equilibrium equations. For instance, if we substitute the divalent calcium ion for the potassium ion in the last system mentioned, the Donnan equations are

$$[Na^+]_1[Cl^-]_1 = [Na^+]_2[Cl^-]_2 \qquad (A\text{-}4)$$

and

$$[Ca^{++}]_1[Cl^-]_1^2 = [Ca^{++}]_2[Cl^-]_2^2 \qquad (A\text{-}5)$$

In order for a calcium ion to be sorbed to the resin, two sodium ions must be simultaneously released from the exchanger. In this case, exchange will take place until the ratio of Na^+ raised to the second power relative to that of Ca^{++} raised only to the first power in both resin phase and solution are equal. This relationship is shown in Eq. A-6.

$$\frac{[Na^+]_1^2}{[Ca^{++}]_1} = \frac{[Na^+]_2^2}{[Ca^{++}]_2} \qquad (A\text{-}6)$$

In recapitulation, the Donnan equilibrium theory explains the following facts that have been observed for either a cation- or anion-exchange resin.

1. Essential exclusion of total free electrolyte (at dilute concentrations) from the commonly used ion-exchange resins is due to the development of a *Donnan membrance potential*. This potential is initiated by the restriction of the nondiffusible exchange sites in the resin particle whose surface is permeable to water and smaller diffusible electrolytes.

2. Ions with a charge sign equal to the exchanger's counter-ions are selectively sorbed from a surrounding solution, while at the same time co-ions (species with a charge sign identical to that on the nondiffusible exchange sites) present in the external electrolyte solution are excluded.

3. The Donnan effect acts independently on each univalent electrolyte species present in a solution; as a consequence, when two or more ionic species are competing for exchange sites in a resin, the concentration ratio of any given pair, at equilibrium, is equal in both the resin phase and in the surrounding solution.

4. The Donnan concept clearly explains the effect of valency in the interaction of polyvalent ions with ion-exchange resins.

5. There can be a substantial reduction of the Donnan effect if a resin is surrounded by a very high concentration of electrolyte. The Donnan effect is also reduced if a resin of low capacity is immersed in an electrolyte whose concentration is high compared to the fixed concentration of the exchanger.

6. The diffusion of nonelectrolytes in and out of an exchanger surrounded by liquid is not influenced by a Donnan membrane potential. Thus not "excluded," uncharged substances tend to establish the same concentration both in the resin particle and in the surrounding solution. This phenomenon is the basis of a separation process known as *ion exclusion* (see Chapter 8).

B. Determination of the Total Ion-Exchange Capacity

Total capacity is defined as the number of exchange-site equivalents per unit weight or volume of the ion-exchange material. Conventional units are: weight capacity, milliequivalents (meq) per gram of dry H-form for cation exchangers and milliequivalents per gram of dry Cl-form for anion exchangers; volume capacity, milliequivalents per milliliter of H-form or Cl-form for cation and anion exchangers, respectively. Lacking label information for ion-exchange materials, capacity may be determined by one of the following methods (see Helfferich* who devotes an entire chapter to capacity definitions, units, and determinations).

**Strong-Acid or Strong-Base Exchangers
That Do Not Undergo Extreme Expansion or Contraction
When Transformed from One Form to Another**

Convert about 15 g of a cation exchanger to the H-form by treatment with excess 1 M HCl using repeated batch or filtration techniques and then wash the exchanger with excess water to remove residual HCl. Next the interstitial water is removed by gentle suction or by centrifuging the sample of exchanger in a closed tube that contains a filter disk. About 10 g of the moist exchanger is weighed and placed in a small column. Excess standard 0.1 N NaOH (about 50 ml per g of exchanger) is passed through the exchanger followed by about five column volumes of rinse water. The effluent including the rinse solution is back-titrated with standard 0.1 N HCl. The water content of the approximately 5 g of exchanger not packed into the column is determined by dehydration techniques (e.g., by drying in vacuum over P_2O_5). The weight capacity is calculated from†

$$\text{Capacity} = \frac{100}{100 - W} \cdot \frac{\text{amount of NaOH} - \text{titrant (meq)}}{\text{weight of moist exchanger (g)}} \qquad \text{(B-1)}$$

where W = water content of the exchanger determined by the dehydration method.

The capacity of a strong-base anion exchanger may be determined in a similar manner. In this case, standard 0.1 N HCl is passed through the OH-

---

*Helfferich, F., *Ion Exchange*, McGraw-Hill Book Company, Inc., New York, 1962, Chap. 4.

†*Ibid.*

form of a known amount of anion exchanger contained in a small column. After the column is rinsed with water, the total eluate is back-titrated with 0.1 N NaOH. The capacity is calculated by an equation analogous to that of B-1.

Volume capacity measurements are made in a similar fashion by noting the packed volume of the fully swollen ion exchanger in the H- or Cl-form as measured in a column before admitting standard solutions to the exchanger bed.

Exchangers that May Expand or Contract Severely When Converted from One Form to Another

Exchangers that fall into this class are the dextrans, celluloses, weak-acid or base types, and the low-crosslinked resinous materials. Capacity measurements on these substances may be carried out by the following general procedures.

(*a*) *Strong-acid or weak-acid types*   Prepare about 5 g of H-form resin and wash with water as before. Place 1 g of weighed material into a dry 250-ml Erlenmeyer flask. Add exactly 200 ml of 0.1 N NaOH in 5% NaCl solution. After shaking and allowing to stand overnight, 50-ml aliquots of supernatant are back-titrated with standard 0.1 N HCl. Water content is determined on the remaining 4 g of exchanger by drying overnight at 60°C under vacuum or at 110°C under atmospheric pressure.

(*b*) *Strong-base or weak-base types*   The procedure is similar to that described in (*a*), except that excess standard 0.1 N HCl is added to the weighed exchanger samples that have been put into the OH-form or free base form. After the neutralization reactions have occurred, the excess standard acid that remains is titrated with 0.1 N NaOH.

C. Determination of the Liquid Volume Held in an Ion-Exchange Column

The total liquid held (the *hold-up volume*) in an exchanger bed consists of the interstitial liquid (that volume between the exchanger particles) plus the liquid contained within the particles themselves. The interstitial or void volume of a column also includes that liquid in any tube or connection leading from the column exit to the point where the effluent is monitored or collected for subsequent assay, and this (the *end volume*) should be excluded from the calculation.

The total void volume (V_0) is determined by passing through the exchanger bed a small amount of sample containing a solute that, because of its size or due to the development of a Donnan membrane potential, is prevented from diffusing into the interior of the exchanger; also, such a solute should not sorb to the outer surface of the exchanger. After correcting for the end volume, V_0 is that volume of eluent required for the nonpenetrating solute to appear in the effluent. Among large-sized solutes employed for V_0 determinations are such polymers as polyphosphates, polyalcohols, polynucleotides, or polysaccharides.[*][†] Among solutes excluded due to a Donnan potential are many simple, easily detected electrolytes such as chlorides, acetates, formates, and iodides. The type of exchanger governs which particular kind of solute should be selected. If a polymer is considered for V_0 determinations, it should be neutral or should have charges identical to those fixed on the exchanger's matrix (i.e., if charged, the polymer should be a co-ion, not a counter-ion). For example ^{32}P-labeled sodium polymetaphosphate was used by Samuelson[‡] to determine V_0 of cation-exchange columns in the Na-form. If a simple electrolyte is considered for V_0 determinations, it should have one of its ions common to the counter-ion of the exchanger and the electrolyte should be used at dilute concentrations. For example, V_0 of a column containing an anion exchanger in the Cl-form may be determined by the following procedure. First, equilibrate the column with dilute sodium chloride solution (e.g., 0.1 M). Second, allow a small volume

---

[*]Rieman III, W., and Walton, H. F., *Ion Exchange in Analytical Chemistry*, Pergamon Press, Inc., New York, 1970, p. 133.

[†]Samuelson, O., *Ion Exchange Separations in Analytical Chemistry*, John Wiley & Sons, Inc., New York, 1963, p. 128.

[‡]*Ibid.*

of 0.1 M hydrochloric acid to drain into the exchanger bed and immediately rinse it through the column with the initially used sodium chloride solution. Third, monitor the effluent for hydrogen ions. V_0 is equivalent to the volume of the sodium chloride rinse solution required for the hydrogen ions to appear in the effluent. V_0 may also be found by a "batch" technique. In this method the exchanger bed is equilibrated with a dilute solution of strong electrolyte of known concentration; one ion of the electrolyte should be identical to the counter-ion of the exchanger. Once the column is equilibrated, the electrolyte solution is drained to the exact top of the exchanger bed, then flushed from the exchanger bed with an excess of water; the amount of electrolyte in the rinse solution is then determined by a suitable analytical technique (e.g., titrimetrically). V_0 is obtained by dividing the amount of electrolyte found in the rinse solution by its original known concentration. Thus, V_0 of a column containing a cation exchanger in the Na-form can be found by titrating, with standard acid, a sodium hydroxide solution (say 0.001 M) that was flushed from the exchanger. For resinous exchangers, V_0 is approximately 38% of the total bed volume (i.e., cross-sectional area $\times$ height).[*][†]

The liquid (V_i) contained inside the exchanger particles may be determined, directly, by noting the weight loss upon drying a known volume of exchanger that has been equilibrated with eluent. First the interstitial liquid is removed by gentle suction or by centrifuging the sample of exchanger in a closed tube that contains a filter disk. Next the moist exchanger is weighed and then following a dehydration step (e.g., drying *in vacuo* at 110°C), the exchanger is reweighed.

Clearly, $V_0 + V_i$ equals the total liquid volume (V_t) contained in the exchanger bed. However, usually V_t is determined as a single indistinguishable volume. Procedures similar to those utilized for measuring V_0 are employed. But in V_t estimations, a solute is required that freely penetrates the exchanger matrix so that it has the same concentration both within the exchanger and in the external solution, i.e., the solute must have a concentration partition coefficient (K_D) equal to unity. Ideally, V_t is found by measuring the volume to the peak maximum of a band of tritiated water after it has emerged from a column. Because of solute-matrix interactions it is often difficult, in practice, to find simple, small molecules that have a K_D of unity. Deviations of K_D from unity for solutes such as formaldehyde, ethylene glycol, glucose, or uncharged nucleosides depend on the type, crosslinkage, and counter-ion form of the exchanger.[†] Formaldehyde (detected easily by the chromotropic acid test) has a K_D very close to unity on most salt forms of general-purpose ion exchangers, and likewise uridine (detected by ultraviolet absorption) distributes evenly internally and externally in an anion

[*]Rieman III, W., and Walton, H.F., *op. cit.*
[†]Wheaton, R. M., and Bauman, W. C., *Ann. N. Y. Acad. Science*, **57**, 159 (1953).

exchanger at pH's below 5 and in a cation exchanger at pH's below 7. In choosing a solute for V_t measurements, one should be certain either by first-hand knowledge or by consulting the literature (e.g., see footnote †) that the solute will have a K_D of unity under a certain set of operating conditions. Obviously, once V_t is determined, V_i may be obtained by subtracting V_0 from V_t.

INDEX

A

Absorption, 13
Absorptivity coefficients, 118
Acid-soluble nucleotides (*see* Nucleoside phosphates)
Acrylic type exchanger, 4
Adenine (*see* Bases, purine)
Adenosine (*see* Nucleosides)
Adenosine diphosphate, 176–177
Adenosine monophosphate, 176–177
Adenosine triphosphate, 176–177
Adsorption, 13
Alcohols, separation of, 216–217
Aldehydes, separation of, 168
Aliphatic carboxylic acids, separation of, 159–162
Alkali metals:
  removal from heavy metals, 136
  separation of, 186–187
Alkaline earths:
  determination of, in sea water, 135
  separation of, 187–193
Amines, separation of, 165–168

Amino acids, 136, 141, 145–148
  automatic analysis of, 145
  isolation from sea water, 136
Amino sugars, 164
  automated procedures for the analysis of, 164
  determination of, in glycoproteins, 164
  separation in the presence of borate, 164
Amphetamine drugs, separation of, 225
Analytical chemical operations as simplified by ion-exchange processes:
catalysis, 132–134
  advantages of ion-exchange method, 132–133
  batch method, 132–133
  column method, 132–133
  by enzymes immobilized on ion-exchange materials, 134
  hydrolysis of various organic compounds, 133–134

243

Analytical chemical operations as
simplified by ion-exchange
processes, catalysis (cont.)
nucleotides, hydrolysis of, 133
sucrose, hydrolysis of, 133
conversion of one compound to
another, 127
by anion exchange, 127
by cation exchange, 127
conditions for, 127
deionized water, preparation of,
129–130
iodate, determination of, following
periodate oxidation of glycol
groups, 131–132
organic compounds, purification of:
by formation of ionized complexes, 130–131
removal of ionic impurities, 130
by sorption to exchangers,
130–131
recovery of trace constituents,
134–141
alkaline earths from sea water,
135
amino acids from sea water, 136
from chromatographic peaks,
140–141
from effluents of industry, 135
of fluoride from potable waters,
135
in food products, 138–139
from fresh or ocean waters, 135
in general environmental work,
135–137
of metals from sea water, 135–137
in the ore and metal industries,
139–140
removal of interfering ions, 128–129
by anion exchange, 129
by cation exchange, 129
standardization of salt solutions, 128
determination of normality, 128
general procedure for, 128

Anion exchangers:
(*see* Dextran ion exchangers)
(*see* Inorganic ion exchangers)
(*see* Cellulosic exchangers)
(*see* Ion-exchange resins)
(*see* Miscellaneous ion-exchange
materials)
(*see* Pellicular ion exchangers)
(*see* Polyacrylamide ion exchangers)
(*see* Selecting the proper ion-
exchange material)
Area under elution curves, 112–119
(*see also* Quantitation of elution
curves)
Aromatic acids, separation of, 162,
219
Aromatic hydrocarbons, separation of,
217
Ascorbic acid, separation of, 162

B

Barium (*see* Alkaline earths)
Bases, purine and pyrimidine:
separation on conventional
exchangers, 171–173
separation by ligand exchange,
226–228
separation on pellicular exchangers,
171–173
Biogenic amines, 167
Bisulfite complex, 168
Body fluids, analysis of, 162, 170
Borate complex:
of amino sugars, 164
of nucleosides, 158–159
of sugars, 154–158
Breakthrough technique, 31–37
breakthrough ion, 35
determination of breakthrough
capacity, 35
determination of optimum chromatographic conditions by, 36

Breakthrough technique (cont.)
  determination of theoretical
    capacity, 35–36
  determination of volume distribu-
    tion coefficient by, 36
  exploratory experiments with, 36–37
  plotting of data from, 35
Bromate, 201
Bromide, 197

C

Calcium (*see* Alkaline earths)
Capacity, determination of, 237–238
  (*see also* Total ion-exchange
    capacity)
Carbohydrates:
  anionic borate complexes of,
    154–159
  automatic analysis of, 155–158
  ionic derivatives of, 158–159
  separation on anion exchangers,
    154–158
Carbohydrates, partition chroma-
    tography of (*see* Partition chro-
    matography of carbohydrates)
Carbonyl compounds, separation of, in
    the presence of bisulfite, 168
Catalysis, ion-exchange method of,
    132–133
Cation exchangers:
  (*see* Dextran ion exchangers)
  (*see* Inorganic ion exchangers)
  (*see* Cellulosic exchangers)
  (*see* Ion-exchange resins)
  (*see* Miscellaneous ion-exchange
    materials)
  (*see* Pellicular ion exchangers)
  (*see* Selecting the proper ion-
    exchange material)
Cellulosic exchangers:
  binding sites of, 59
  crosslinking with bifunctional
    reagents, 69

crystalline regions of, 67–68
dense particles of, 69
hydrogen bonding in, 67, 69
hydrophilic nature of, 59
ionizable groups of, 67
matrix of, 67
multiple bonding of, 59
pore size, 59, 67, 69
preparation of, 67–68
separation of macromolecules on,
  59, 68–69
surface area of, 59
Chelating resins:
  in ligand-exchange reactions,
    224–230
  recovery of trace constituents by,
    136
  select affinity of, 80
  structure for, 80
Chlorate, 201
Chloride (*see* Halides, separation of)
Chromatographic assembly, operation
    of, 108–109
Chromatographic columns (*see also*
    Laboratory columns and
    accessories)
  bed density of packing, 28
  elution analysis with, 37–42
  fabrication from glass, 89
  homemade types, 89
  importance of shape, 87
  optimal conditions in, 36–39, 53–54
  plates, determination of number in,
    47–49
  professional designs for, 89
  ratio of length to diameter, 87
Chromatographic nomenclature:
  definitions, 25–27
  symbols, 25–27
  units, 25–27
Chromatography, types of:
  ion-exchange, 144–201
  ion-exclusion, 222–223
  ligand-exchange, 224–230

Chromatography, types of (cont.)
  partition, 208, 211–215
  salting-out, 216–218
Co-ions:
  complex formation with, 12
  definition of, 12
  exclusion from exchanger matrix,
    208, 235
Column efficiency (*see also*
    Resolution)
  and degree of resolution, 123
  as measured by HETP, 124
  in terms of number of theoretical
    plates, 123
Column volumes, 27, 37, 44–51
  relation to distribution ratio, 44
  relation to elution volume, 37
  in terms of geometrical dimensions,
    27, 49
  in terms of number of void volumes,
    44–46
Columns, packing of, 87, 89–94
    (*see also* Chromatographic
    columns)
  choice of procedures for, 93–94
  imperfections in packing of, 87
  pretreatments of packing materials,
    89
  techniques for:
    multistage batch-packing, 93
    single-stage batch-packing, 93
    single-stage pump-packing, 93
Complexing agents, separation of
    metals in the presence of,
    182–186, 192–197 (*see also* Ion-
    exchange chromatography of
    inorganic substances)
Complex-ion interactions (*see* Ion
    exchange – complex ion
    interactions)
Constancy of operation conditions:
  importance for quantitative results,
    119

use of internal standards to check
    on, 119–120
Conversion of one compound to
    another, 127
Copolymerization of styrene and
    divinylbenzene, 2–3
Counter-ions:
  affinity of, 11–12
  defined, 8
  in Donnan equilibria, 13–16
  equilibria of, 12
  in identifying the exchanger, 64
  ionic bonds of, 12
  in ligand-exchange reactions, 224
  penetration into exchange matrix,
    208, 235
  polarizability of, 12
  in reversible reactions, 17–18
  selectivity of resin for, 10–11, 19–21
Crosslinkage, 8–9, 65, 69, 72–74
Cytidine (*see* Nucleosides)
Cytosine (*see* Bases, purine and
    pyrimidine)

D

Deionized water, 129
Deoxynucleotides, 141
Detection of solutes in the effluent,
    102–104
  analytical methods for, 102
  automatic devices for, 102
  detector response signals for, 102
  flow through cells for, 102
Dextran ion exchangers (*see also*
    Selecting the proper ion-
    exchange material)
  binding sites, 59
  chemical and physical properties,
    74–75
  crosslinkage of, 75
  functional groups of, 74
  hydration of, 74

Dextran ion exchangers (cont.)
hydrophilic nature of, 59, 72
multiple bonding of, 59
porosity, 75–76
preparation of, 72
separation of low molecular weight
substances on, 76
separation of macromolecules on,
59, 75–76
sieve effect on, 75
as three-dimensional copolymer, 71
Diffusion (*see* Rate of diffusion)
Displacement development, 31–34
breakthrough point in, 33
cross-contamination in, 33
requirements for, 31
separation of Na, K, and Cs by, 33
sharp boundaries or elution bands in,
33
Distribution coefficient, volume (D_v):
defined, 26, 36
degree of separation as indicated by
ratios of, 37
equations for, 29, 36
relation to number of void volumes,
44
Distribution coefficient, weight (D_g),
21–22, 26, 28–29
defined, 21
equation relating to column operations, 28–29
equations for, 28
ratio of, 22
relation to chromatographic
separation, 21–22
relation to elution volume, 22
relation to separation factor, 22
Divinylbenzene, 2
crosslinks, 2–3
polymers, 2–3
resins, 2–3
DNA, separation of, 222–223

Donnan membrane theory:
as applied to ion-exchange
reactions, 15–16
condensed summary of, 236
diagrammatic representation of, 14
effect of valency in, 235–236
electrolyte-exchanger interactions
of, 208–211, 235–236
exclusion phenomenon of, 209,
235–236
and mixed electrolyte systems,
235–236
and polyvalent ions, 235–236

E

EDTA (ethylenediaminetetraacetic
acid), 138, 195
Effluent volume (*see* Elution volume
or "ml to peak")
Electrolytes:
diffusion of, 208–210
exclusion of, 209
partition of, 208–210
Elution (*see also* Elution analysis)
gradient, 97, 99, 101
stepwise, 98, 101
Elution analysis:
description of, 37–39
elution curves of, 39
elution volume of, 39–40
equation for, 40
relation to number of column
volumes, 39
relation to volume distribution
coefficient, 39–40
features of method, 37–38
plate theory of, 42
resolving power of, 40
solute bands in:
diffusing boundaries of, 39
distance between peak maximum,
121–122

Elution analysis, solute bands
in (cont.)
Gaussian shape of, 39, 121
maximum concentration of, 39, 44
plates encountered by, 53
resolution of, 121
width of, 39, 53, 121–122
Elution curves (*see also* Elution analysis)
area, methods of obtaining, 112–119
automatically plotted, 112
band widths of, 121–122
calculation of theoretical shapes for, 46–47, 51
degree of overlap, 49, 52, 121
degree of resolution, 49, 121
detected spectrophotometrically, 118
experimental, 54
impurity ratios for, 52–53
parameters affecting shape of, 122–124
peak maxima of, 121
quantitation of, 111
relation to sample composition, 115
relation of solute content to area under, 115–119
shapes of, 39, 54
spreading of, 121
variation of height and width of, 49, 51, 53
ways of plotting, 111
Elution maximum (*see* Elution analysis)
Elution volume or "ml to peak" (*see also* Elution analysis)
equation for, 27, 37
maximum concentration of solute in, 39–40
relation to column dimension, 37
relation to number of column volumes, 40
Enzymes, immobilized on polymeric supports, 134

Equilibrium quotient, 17–18
relation to selectivity coefficient, 18
units for, 18
usefulness of, 18–21

F

Fluoride, 129, 135
Food products, isolation of trace elements in, 138–139
Fraction collectors, 105
Frontal development, 30–31
description of, 30
separation of mixtures by, 30–31

G

Galactosamine, 164
Galacturonic acid, 160
Gaussian shapes:
curve of error, 39
elution curve and relation to error function, 46
elution zones, 39–40, 121
and Mayer-Tompkins equation, 44–46
Glucosamine, 164
Glucuronic acid, 160
Gradient elution, 96–101, 169
as a continuous process, 98
defined, 96–97
devices for, 99
effectiveness of, 169
equation for rise in eluent power by, 99
objective of, 100
schematically illustrated advantages of, 100
Guanine (*see* Bases, purine and pyrimidine)
Guanosine (*see* Nucleosides)

H

Halate ions, separation of, 201
Halides, separation of, 197–199
  on hydrous zirconium oxide,
    197–199
  on resinous exchanger, 197–199
HETP (height equivalent to theoretical
    plate), 42 (*see also* Plate theory
    of chromatography)
Hexosamines, 164
Hold-up volume, 239–240
Hydrated ions, 8, 11
Hydrogen bonding, 179–181
Hydrolysis, using ion-exchange
    materials, 132–133

I

Inorganic anions, separation of,
    197–201
Inorganic cations, separation of,
    186–197
Inorganic ion exchangers (*see also*
    Selecting the proper ion-
    exchange material)
  acid salt exchangers:
    cation behavior of, 78
    separation of alkali and alkaline
      earth ions on, 78
    strong affinity for cesium for, 78
    zirconium phosphate as typical
      example of, 78
  description, preparation, and
    properties of, 60, 76–80
  heteropoly acid salt exchangers:
    crystal lattice of, 79
    examples of, 79
    reaction with multivalent cations,
      79
    strong affinity of alkali metals
      toward, 79

hydrous oxide exchangers:
  behavior as either cation or anion
    exchangers, 77
  charge on matrix of, 77
  high affinity for polyvalent anions,
    77
  prepared from tetravalent metals,
    77
  unique attraction for fluoride ions,
    77
inorganic phosphate gels:
  calcium phosphate gel as example
    of, 80
  purification of macromolecules
    on, 80
  uses of in biochemical work, 80
metal sulfide exchangers:
  as absorbents for the transition
    metals, 79
  examples of, 79–80
  removal of trace components with,
    80
  sorption-displacement type
    reaction of, 79
  unique affinity toward certain ions,
    60, 76–79
  unique properties of, 76
Interfering ions, removal of, 128
Internal standards, 119–120
  to check operating conditions, 119
  to normalize chromatograms, 120
  requirement for, 120
Interstitial volume, 16, 36, 40, 49,
    208–210, 239–241 (*see also*
    Void volume)
Iodate:
  determination of, 131
  separation from other halates, 201
Iodide (*see* Halides, separation of)
Ion-exchange beads:
  cross-section view of, 7–8
  diffusion in, 7, 10, 13, 64
  gel-like structure, 7, 13

Ion-exchange beads (cont.)
  particle size, 6, 63
  water uptake, 7–8
Ion-exchange chromatography of
    inorganic substances, 186–201
    (*see also* individual listings of
    compounds)
  inorganic anions, 197–201
    halate ions, 201
    halide ions, 197–199
    phosphorus anions, 200–201
    sulfur compounds, 201
  inorganic cations, 186–197
    alkali metals, 186–187
    alkaline earth ions, 187–192
    miscellaneous metals, 197
    rare earth elements, 195–196
    transition elements, 193–194
Ion-exchange chromatography of
    organic substances, 145–181
    (*see also* individual listings of
    compounds)
  amines, 165–168
  amino acids, peptides and proteins,
    145–154
  amino sugars, 164
  carbohydrates, 154–159
  carbonyl compounds, 168
  nucleic acid components, 168–181
  organic acids, 159–162
  simultaneous analysis of different
    classes of compounds,
    181–182
Ion exchange – complex ion inter-
    actions, 183–187, 193–197
  of alkaline earths, 183–184,
    187–193
  behavior of metals in, 183–187,
    193–197
  broad survey of, 184
  on cellulosic exchangers, 197
  complexing agents for, 183–197
  on inorganic exchangers, 186
  of metal formates, 184
  of rare earths, 183–184
  on resinous exchangers, 186
Ion-exchange elution chromatography,
    182–183 (*see also* Elution
    analysis)
  complexing agents in, 183–186
  emergence of, 182
  mechanisms of, 183
Ion-exchange equilibria, 12–13, 63,
    235–236
  described by Donnan theory, 13, 17,
    235–236
  described by mass-action law, 17
  description of, 12–13
  effect of particle size on, 63
Ion-exchange literature, 81–83 (*see
    also* Selecting the proper ion-
    exchange material)
Ion-exchange materials:
  classification of, 2
  different types of, 2–4
  nature of, 2
  versatility of, 1
Ion-exchange processes, 25–54
  batch method, 27–29
    cascade operation, 28
    relation to column operation, 28
    single-stage operation, 27–28
  column method, 29–42
    breakthrough technique, 31–34
    displacement development, 31–34
    elution analysis, 37–42
    frontal development, 30
    multistage operation of, 28
    the plate theory of, 42
    terminology of, 25–27 (*see also*
      Chromatographic nomen-
      clature)
Ion-exchange reactions:
  application of Donnan membrane
    theory, 12–17, 235–236
  application of law of mass action, 17
  equations for, 6
  equivalency of, 9–10

Ion-exchange reactions (cont.)
mechanisms of, 12–13
rate of, 22–23
stoichiometry of, 9
Ion-exchange resins:
affinity, 11
of counter-ions for, 11
order of, 11
anion-exchanger, types of, 2–3, 5–6
defined, 2
preparation of, 2–3
structures for, 3, 5–6
cation exchanger, types of, 2–5
defined, 2
preparation of, 2–3
structures for, 3, 5
chemical behavior of, 3
chemical formulas for, 5–6
chemical properties of, 9–12, 65–66
classes of, 2, 61
degree of crosslinkage, 8–9, 64
functional groups of, 2–3, 8, 59, 61
hydration of, 8, 58
hydrocarbon skeleton as part of
structure of, 2–3, 8
immersed in electrolyte solution, 2,
10, 13–14, 63–64, 235
interstitial volume of, 16
ionic forms of, 64–65
matrix or lattice material of, 2–3,
10, 58–59
mobile (active) ions of, 2
particle size of, 63
for batch work, 63
for column operation, 63
physical properties of, 6–7
pore liquid, 8
pore size, 8, 58–59, 64
porosity and swelling, 8, 58, 64
selection of (*see* Selecting the proper
ion-exchange material)
selectivity, 10–11, 18–21, 65–66
coefficients, 11, 18, 20–21
determination of, 21

factors that contribute to, 11–12
practical value of, 18, 21
scales, 11, 18–21
sequences, 11
symbols for, 5–6, 65–66
utilization of, for partition, salting-
out and ion-exclusion chromatography, 208–223
Ion-exchangers:
capacity of, 237
cellulosic (*see* Cellulosic exchangers)
chart to choose from, 57
dextran (*see* Dextran ion exchangers)
inorganic (*see* Inorganic ion
exchangers)
materials used as, 2
miscellaneous ion-exchange
materials, 80
multiple bonding of, 59
pellicular (*see* Pellicular ion
exchangers)
polyacrylamide (*see* Polyacrylamide
ion exchangers)
resin type (*see* Ion-exchange resins)
selection of, 56–83 (*see also*
Selecting the proper ion-
exchange material)
Ion-exclusion, 16, 208–223, 235–236
(*see also* Ion-exclusion
chromatography)
defined, 209
and Donnan theory, 208, 211
of ionized compounds, 218–223
repulsive forces of, 218–222
in separation of electrolytes from
nonelectrolytes, 208–211
separation of ionic compounds by,
218–223
Ion-exclusion chromatography, 17,
219–223 (*see also* Ion-exclusion)
on anion exchangers, 222
on cation exchangers, 220–222
of nucleic acid compounds, 222
of nucleosides, 222

Ionic compounds, separation of:
  by ion-exclusion, 218–223
  by sorption chromatography,
    219–223
Ionic radius, 11
Ionic sieve effect, 58–59

K

Krebs-cycle acids, separation of, 159,
    162

L

Laboratory columns and accessories,
    86–109
  apparatus for delivering eluent to
    the columns, 96
    gradient elution reservoirs, 99
    single-solvent reservoirs, 99
  chromatographic assemblies, 86
    block diagram for, 86
    operation of, 108
  columns:
    designs of, 88–89
    length to diameter ratio, 87
    materials used for, 88–89
    packing of, 89–94
  detection devices, 102–103
    classification of, 103
    flow through cells of, 102
    sensitivity of, 102–103
  devices to control rate of eluent
    flow, 104–105
    flow by gravity, 104
    mechanical pumps, 104–105
    pressurized vessels, 104
  fraction collectors, 105, 107
    automatic, kinds of, 107
    commercial models, 107
  line connections between
    accessories, 95–96
    for high-pressure systems, 95
    for low-pressure systems, 95

sample loading, 108–109
  large volumes, 109
  small volumes, 109
  via injection ports or valves, 109
  via loop-type injector, 109
Law of chemical equilibrium, 17
Ligand exchange:
  coordinating metals of, 224
  counter-ions of, 224
  defined, 224
  equation for, 224
  as a highly selective process, 225
  ligand molecules, 224–225
  resin types for, 224–225
Ligand-exchange chromatography:
  amphetamine drugs, separation by,
    225
  on chelating resins, 224–230
  peptides and amino acids, separation
    by, 228–230
  purines and pyrimidines, separation
    by, 225–226
Liquid flow through the column, 104
Liquid held in ion-exchange column,
    239–241
  defined, 239
  determination of, 239–241
Literature, ion-exchange, 81–83 (*see
    also* Selecting the proper ion-
    exchange material)

M

Magnesium (*see* Alkaline earths)
Mayer-Tompkins equation (*see* Plate
    theory of chromatography)
Metal industries, ion-exchange
    processes of, 139–140
Metals, separation of (*see also* listings
    for individual compounds)
  miscellaneous types, 197
  in presence of complexing agents,
    182–186, 192–197

Miscellaneous ion-exchange materials:
chelating (iminodiacetic acid) type,
80
intermediate acidic or basic types,
81

N

Nonchromatographic operations, 126
Nonelectrolytes:
partition coefficients of, 210-211
penetration of, into exchanger
phase, 208-211
separation of, 209-211
Nucleic acid components, separation
of, 168-181 (*see also* Bases,
purine and pyrimidine; Nucleoside phosphates, separation of;
Nucleosides; Nucleotides)
automated analyzer, 168-170
historical account, 168-169
modern trends of, 169-171
nucleic acids, 181
nucleosides, 174
nucleoside phosphates, 175-179
nucleotides, 175-179
polynucleotides, 179-181
purine and pyrimidine bases,
171-174
Nucleic acids, 181
Nucleoside phosphates, separation of:
acid soluble (tissue nucleotides),
177-179
on conventional resins, 177
dinucleoside monophosphates, 177
guanosine, mono-, di-, tri-, and
tetraphosphates, 176
nucleotides (mono):
deoxyribonucleotides, 175-176
ribonucleotides, 175-176
on pellicular exchangers, 177-179
ribonucleoside 5'-mono-, di-, and
triphosphates, 177-179

Nucleosides:
borate complex of, 158-159
dinucleoside monophosphates, 177
separation of, 158-159, 174
separation of, by ion-exclusion
chromatography, 219-222
Nucleotides (*see also* Nucleoside
phosphates, separation of)
borate complex of, 158-159
dioxyribo-, 175-177
poly-, 179-181
ribo-, 175-176
separation of, 158-159, 175-179

O

Organic acids, 159-162
aliphatic carboxylic acids, separation
of, 160-161
aromatic acids, separation of, 162
automatic analysis of, 159-162
detection of, in column eluates,
159-160
Krebs-cycle acids, separation of,
159, 162
uronic and aldonic acids, separation
of, 160-161
Organic compounds:
purification of, 130
separation of, 145-182

P

Packing of chromatographic columns
(*see* Columns, packing of)
Partition chromatography of carbohydrates, 211-215
on anion exchangers, 211-214
automation of, 211-215
on cation exchanger, 215
with organic bases as counter-ions,
214
Peak area, 112-119

Peak elution volume (*see* Elution volume or "ml to peak")

Pellicular ion exchangers:
nature of, 81
rate of diffusion in, 81
review on, 81
superficially porous crust of, 81

Peptides and amino acids:
separation by ligand exchange, 228–230

Peptides and proteins, 148–154
separation on cellulosic exchangers, 151–153
separation on resinous exchangers, 148–151

Periodate oxidations, 131

Phenol derivatives, separation of, 218–219

Phosphate polymers, separation of, 199–201

Phosphorus anions, separation of, 200–201

Plate theory of chromatography:
column efficiency, 123
column performance in relation to:
dependency upon equilibrium, 53
factors effecting plate height, 53
optimal conditions, 43, 54
conditions of, 43, 53
continuous approach, 50–51
calculation of theoretical elution curve, 51
description of, 50
mathematical treatment of, 50–51
terminology for, 50
discontinuous approach, 42–50
calculation of theoretical elution curve, 44–46
definitions, notations, and symbols for, 43
description of, 43–44
mathematical treatment of, 42–50
significant outcome of, 44

the elution curve as a Gaussian error function, 46–47, 51
calculation of theoretical elution curve, 47, 51
derivations of equations for, 46–48
normalization of Mayer-Tompkins equation, 46

"plate"
calculation of, from elution curve, 123
definition of, 42
effect of particle size on height of, 63
(HETP) height equivalent for theoretical, 42, 53, 123–124
methods for determining number of, 47–49
relation to column length, 53, 123–124
predictions of, 49

Polyacrylamide ion exchangers:
binding sites of, 59
chemical and physical properties of, 74–75
crosslinkage of, 72–76
functional groups of, 73–74
hydration of, 74–75
hydrophilic nature of, 59, 72–73
ionogenic groups of, 71
pore size, 59, 75–76
preparation of, 72
separation of low molecular weight substances on, 76
separation of macromolecules on, 59, 75–76
sieve effect of, 75

Polynucleotides, separation of, 179–181
homologous oligonucleotides, 179
mixed polynucleotides, 179–181
oligodeoxyadenylates, 179
oligomers, 179
in the presence of urea, 179

Potassium (*see* Alkali metals)
Proteins (*see* Peptides and proteins)
Pumps, types of, 104–105

Q

Quantitation of elution curves, 111–120
  automatically plotted, 112
  determination of area under,
    112–119
    geometrical integration of, 113
    planimeter tracing of, 114–115
    weighing of, 114–115
  constancy of operating conditions,
    dependency for quantitative
    results, 119–120
    use of internal standards to check
      on, 119–120
    use of standard chromatogram to
      check on, 119–120
  conversion of peak area to amount
    of solute, 115–119
    response factors for, 115, 117
    from standard chromatogram, 115
  conversion of peak area to amount
    of solute by use of absorptivity
    coefficients, 118–119
  from peak height – peak width
    measurements, 113–114
  relation to sample composition, 115
  individual fractions, 111–112
    analysis of, 111
    plots of elution curves for, 111–112
  integration methods, precision and
    accuracy of, 120

Rate of diffusion, 7, 10, 12–13, 53,
  58, 64, 81, 209
Rate of ion-exchange reactions, 22–23
  kinetics, 22–23
    film diffusion, 22–23
    particle diffusion, 22–23
    steps involved in, 23
Resolution, 49, 53, 121–124
  consistent units for, 121
  by decreasing width of elution
    curves, 123–124
  equation for, 121
  by increasing column length,
    122–124
  by increasing distance between
    peak maxima, 122–124
  quantitative measure of, 121
  relation to plate theory predictions,
    121
  relation to separation factor and
    peak elution volumes, 121
  theoretical considerations of,
    121–124
Resolution factor, 37, 121–124
  indication of degree separation by,
    37, 53
Response factors (*see also* Quantita-
    tion of elution curves)
  calculation of, 115–117
  relation to area under elution curves,
    115–117
  relation to height of elution curves,
    117
Retention time, 119
Ribose phosphates, 141
Rubidium (*see* Alkali metals)

R

Radioactive wastes, disposal of, 138
Rare earths, separation of, 195–197
  by anion exchange, 195–196
  as EDTA complexes, 195–197

S

Salting-out chromatography, 216–218
  of aliphatic alcohols, 216–217
  of aromatic hydrocarbons, 217
  defined, 216

Salting-out chromatography (cont.)
  nonionic substance, separation by,
    216–217
  of soap samples, 217–218
Sample loading devices, 109
Selecting the proper ion-exchange
    material:
  by elimination process:
    based on the chemical and physical
      environment of a solute, 60
    based on the net charge of a solute,
      58
    based on the size and net charge
      of a solute, 57
  inorganic ion exchangers: (*see also*
      Inorganic ion exchangers)
    acid salt types, 78
    heterapoly acid salt types of, 79
    hydrous oxides of tetravalent
      metals as, 77
    inorganic phosphate gel as, 80
    metal sulfides as, 79–80
    rebirth of, 76
    review on, 78
    special uses of, 76
    unique selectivities of, 77–78
  ion-exchange celluloses (*see also*
      Cellulosic exchangers):
    choice of exchanger type, 69
    descriptions of, 67–68
    fibrous, 67
    microgranular, 68
    table of, 70
  ion-exchange dextran (*see also*
      Dextran ion exchangers)
    gel matrix of, 71
    general characteristics of, 74–75
    selection of exchanger type, 75–76
    table of, 72
  ion-exchange literature, 81–83
    books, 81–82
    handbooks, 82–83
    journals, 83
    reviews, 83

  ion-exchange polyacrylamides (*see
      also* Polyacrylamide ion
      exchangers)
    gel matrix, 71
    general characteristics of, 74–75
    selection of exchanger type, 75–76
  ion-exchange resins (*see also* Ion-
      exchange resins)
    chemical and physical stability, 65
    classification of, 61
    formulary for identifying, 65
    four major types of, 61
    ionic form of, 64
    per cent crosslinkage and capacity,
      64
    table of, 62
    terminology, 61–65
    trade names, manufacturers, and
      suppliers, 60–61
  miscellaneous ion-exchange mate-
      rials, 80
    chelating type, 80
    intermediate acid type, 81
    intermediate basic type, 81
    pellicular ion exchangers, 81
    select affinities of, 80
Selectivity coefficients, 11, 18, 20–21,
    38
Separation (*see* Resolution)
Separation of:
  electrolytes from non-electrolytes,
    208–211
  inorganic compounds, 186–201
  ionic compounds by ion-exclusion,
    218–223
  nonelectrolytes, 210–211
  organic compounds, 210–211,
    218–222
Separation factor, 22, 121
  definition of, 22
  usefulness of, 22
Simultaneous analysis of different
    classes of compounds, 181–182
Sodium (*see* Alkali metals)

Solutes, penetration into exchange matrix:
co-ions, 208, 235
counter-ions, 208, 236
for determining void volume and total "hold up" volume, 239-241
electrolytes, 208-209, 235-236
factors effecting, 208-210
interactions with matrix, 210
uncharged molecules, 209-211
volume partition coefficients of, 211
Sorption, 13
Sorption chromatography, 219
Standardization of salt solutions, 128
Start-up procedure for column operation, 108
Stepwise elution, 96-101
devices for, 99
as a discontinuous process, 98
Strontium (*see* Alkaline earths)
Sugar phosphates, 158-159
Sugars (*see also* Carbohydrates)
amino, 164
ionized derivatives of, 158-159
neutral, 154-158
Sugars, partition chromatography of (*see* Partition chromatography of carbohydrates)
Sulfur compounds, separation of, 201

T

Total ion-exchange capacity:
defined, 237
method for determining, 237-238
Total salt concentration, determination of, 128

Trace constituents, recovery of, 134-141
Transition elements:
removal from sea water, 136
separation of, 193-194

U

Uncharged solutes, 16, 209-211, 236
Uracil (*see* Bases, purine and pyrimidine)
Uridine (*see* Nucleosides)
Uronic and aldonic acids, separation of, 160-161

V

van der Waals forces, 12, 16, 218
Void volume, 208-210 (*see also* Interstitial volume)
in fundamental equation of partition chromatography, 210
in ion exclusion, 209
total "hold-up" volume, 239-241
Volume partition coefficients, 210

W

Water, total free, contained in exchangers, 209-210
"Wrong way" chromatography, 217, 219-223
separation of aromatic compounds by, 217
separation of DNA by, 222-223
separation of nucleosides by, 219-222